T0321019

TAILOR MADE CONCRETE STRUCTURES
NEW SOLUTIONS FOR OUR SOCIETY

PROCEEDINGS OF THE INTERNATIONAL *FIB* SYMPOSIUM 2008, AMSTERDAM, THE NETHERLANDS, 19–21 MAY 2008

Tailor Made Concrete Structures: New Solutions For Our Society

Edited by

Joost C. Walraven
Technical University of Delft, Delft, The Netherlands

Dick Stoelhorst
Dutch Concrete Society, Gouda, The Netherlands

CRC Press
Taylor & Francis Group
Boca Raton   London   New York   Leiden

CRC Press is an imprint of the
Taylor & Francis Group, an **informa** business

A BALKEMA BOOK

CRC Press/Balkema is an imprint of the Taylor & Francis Group, an informa business

© 2008 Taylor & Francis Group, London, UK

Typeset by Charon Tec Ltd (A Macmillan Company), Chennai, India
Printed and bound in Great Britain by Antony Rowe (A CPI-group Company), Chippenham, Wiltshire

Published by: CRC Press/Balkema
                P.O. Box 447, 2300 AK Leiden, The Netherlands
                e-mail: Pub.NL@taylorandfrancis.com
                www.crcpress.com – www.taylorandfrancis.co.uk – www.balkema.nl

ISBN 13: 978-0-415-47535-8 (Hardback + CD-ROM).

Table of Contents

Oral Presentations

Life cycle design

Design strategies for the future

Codes for the future

IX

X

Developing a modern infrastructure

Designing structures against extreme loads

Increasing the speed of construction

Poster Presentations

Life cycle design

Design strategies for the future

Monitoring and inspection

Diagnosis

Innovative materials

Codes for the future

Increasing the speed of construction

Tailor Made Concrete Structures – Walraven & Stoelhorst (eds)
© 2008 Taylor & Francis Group, London, ISBN 978-0-415-47535-8

Preface

Since the last FIP congress 10 years have gone by, a lot has happened in the new organisation *fib*.

The successful Congresses in 2002 and 2006 in Osaka and Naples, many technical publications, the preparations for a new model code, a lively set of Commissions, Task Groups and Special Activity Groups, all show *fib* is alive and kicking.

Therefore organising the official 2008 *fib* symposium, 10 years after the merger of FIP and CEB during the FIP Congress in 1998 in Amsterdam, is a very special occasion. It gave the organisers, the sponsors a warm feeling about the development of *fib*.

When we set up the organisation of the event we were overwhelmed by the support we got from companies in The Netherlands. Authorities, Contractors, Suppliers, Consultants, etc. helped us at a very early stage, so we could start off. Once again we showed we are interested in contacts with our colleagues all over the world. And the result was great; the response to our request for Call for Papers resulted in a total number of more than 390 abstracts from all over the world. We have to thank all authors for their support to our event.

The Scientific Committee had a tough job to review all these abstracts. We are very grateful for this effort. And now you see the final result; the proceedings of the symposium and a symposium programme we can be proud of.

Many new papers about innovation, new design rules and design strategies, architectural developments in concrete construction; it is only a snatch of the total offer you will find in the programme. Once again it shows Concrete is Developing; *fib* is developing too.

Dick Stoelhorst,
Secretary *fib* 2008 Symposium Amsterdam.

Tailor Made Concrete Structures – Walraven & Stoelhorst (eds)
© 2008 Taylor & Francis Group, London, ISBN 978-0-415-47535-8

Sponsors

Ministerie van Verkeer en Waterstaat

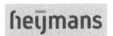

Keynote Lectures

Tailor Made Concrete Structures – Walraven & Stoelhorst (eds)
© 2008 Taylor & Francis Group, London, ISBN 978-0-415-47535-8

Concrete bridges: New demands and solutions

Jacques Combault
President of IABSE; Technical Director of Finley Engineering Group, USA

CURRICULUM VITAE

Jacques is one of the world's leading experts in the design and construction engineering of cable-stayed and long-span bridges. During his distinguished, award-winning career, Jacques played a major role in some of the most innovative bridge projects of the past several decades, including the Sunshine Skyway Bridge in Tampa, the Brotonne Bridge on the Seine River in France, the Rion-Antirion in the Gulf of Corinth, Greece, and the Sutong Bridge Project in China. A frequent collaborator with legendary bridge designer Jean Muller, Jacques is a two-time nominee as *ENR* magazine's Man of the Year.

His long list of honors and recognitions also includes the Federation International du Beton Medal of Merit Award in 2004, the Award of L'Association Francaise pour la Construction in 1991, and three Innovation Awards from Groupe GTM. He also served as vice president of the International Association for Bridges and Structural Engineering (IABSE) and has been a keynote speaker at many industry conferences and meetings.

In addition to serving in key design and technical advisory roles on some of the most complex bridge projects in the world, Jacques is recognized for developing far-reaching innovations in the fields of pre-fabricated concrete and steel-concrete composite bridges. He has a master's degree in engineering from the Ecole Centrale de Lyon, spent many years as a professor at the Ecole Nationale des Ponts et Chaussées in the field of construction techniques, and has published more than 30 scientific and technical papers.

Tailor Made Concrete Structures – Walraven & Stoelhorst (eds)
© 2008 Taylor & Francis Group, London, ISBN 978-0-415-47535-8

How The Netherlands survive, a world wide example

Sybe Schaap
Chairman of the Association of Water Boards, The Netherlands

CURRICULUM VITAE

Family name:	Schaap
First name:	Sybe
Title:	Dr. Ing.
Date of birth:	20-05-1946
Nationality:	Netherlands
Education:	Agricultural College Leeuwarden Economy/Social Science/Philosophy Free University Amsterdam Dissertation Philosophy Amsterdam Habilitation Philosophy Prague
Membership of professional bodies:	President of the Dutch Union of Water boards
Present position:	Senator Chairman Water board Groot Salland Lecturer Philosophy Amsterdam/Prague, Czech Republic Director Consultancy Companies Netherlands/Eastern Europe

Key qualifications:

Main position of Mr. Schaap since 1986: Chairman of Water boards in polder areas in central Netherlands. Tasks: protection against water floods and conditioning of the quantity and the quality of surface water. The water conditioning task is highly integrated with agricultural demands and the development of nature. After being a farmer in the Netherlands for a few years, Mr. Schaap started agricultural consultancy activities all over Europe, in recent years combined with agricultural production in Ukraine. The consultancy activities include the wide range of water management in central Europe and third world regions.

In May 2007 Mr. Schaap was elected Senator of the Royal Kingdom of the Netherlands for the Liberal Party VVD.

Note: Powerpoint presentation by this author can be found in the CD-ROM.

Tailor Made Concrete Structures – Walraven & Stoelhorst (eds)
© 2008 Taylor & Francis Group, London, ISBN 978-0-415-47535-8

The World Champions Soccer in South Africa in 2010

Danny Jordaan
CEO 2010 FIFA World Cup Organisation South Africa

CURRICULUM VITAE

Daniel Alexander "Danny" Jordaan (born September 03, 1951) is a South African sports administrator as well as a former lecturer, politician and anti-apartheid activist. He is best known for leading South Africa's successful Football World Cup 2010 bid.

Born in Port Elizabeth, a city on the southeast coast of South Africa, Jordaan got involved in anti-apartheid activities by joining the South African Students' Organisation (SASO) in the early 1970s. This organisation was founded by Steve Biko in order to defend the rights of black students. Later, Jordaan also became a member of the United Democratic Front and the African National Congress (ANC).

Following his studies, Jordaan became a teacher in 1974. From 1970 to 1983 he was a provincial cricket and football player. In the latter sport, he achieved professional status for a brief period. His political and sport interests soon combined and he became an activist in various organisations fighting to break down racial barriers in sport.

From 1983 to 1992 he served as the president or vice-president of various soccer boards. In 1993 he was appointed as a director of the Cape Town Olympic Bid Company.

His political career also progressed; in 1990 he was elected as the chairperson of the ANC branch in Port Elizabeth North. After the first fully inclusive South African elections in 1994, he became a member of parliament for the ANC, a position he held until 1997.

In 1997, he was elected as the chief executive officer of the South African Football Association (SAFA). He subsequently headed South Africa's unsuccessful Football World Cup 2006 bid, gaining great respect internationally for his work. As a consequence, he also led South Africa's Football World Cup 2010 bid, this time successfully.

Jordaan has served on the marketing and television board of FIFA since 1998. He received a special presidential award from President Nelson Mandela in 1994 as well as the presidential sport achievement award from President Thabo Mbeki in 2001. He won South Africa's marketing person of the year award in 2000.

Jordaan has a BA Honours degree from the University of South Africa.

In 2004, he was voted 44th in the Top 100 Great South Africans.

Tailor Made Concrete Structures – Walraven & Stoelhorst (eds)
© 2008 Taylor & Francis Group, London, ISBN 978-0-415-47535-8

Dam safety in China and the life span evaluation of old concrete dams

Jinsheng Jia & Cuiying Zheng
China Institute of Water Resources and Hydropower Research;
Chinese National Committee on Large Dams

Chun Zhao
China Institute of Water Resources and Hydropower Research

ABSTRACT: More and more old dams are operated for more than 50 years. Evaluation on the life span and the real safety status becomes a challenging task for the dam society, especially for China because of more than 6000 dams to be evaluated and rehabilitated within the next 3 years. Based on the investigation on FENGMAN gravity dam, which is 91.7 m high, operated from 1943 and suffered too much up lift pressure, freeze and thaw problem, etc., discussions on the life span evaluation of the old concrete gravity dams have been made. The reasonable coefficient of dam safety has been discussed and rehabilitation schemes have been recommended. Meaningful results have been achieved based on the case study.

Keywords: Dam safety, Life span, Rehabilitation.

1 LIFE SPANS OF OLD CONCRETE DAMS AND DAM SAFETY IN CHINA

For concrete gravity dams built since the 20th century, there are no standards for the normal working period (life span) in China or other countries. According to literatures, a large number of dams lower than 30 meters were built 1000 years ago, but few of them exist nowadays. Most of these dams have failed and their life span was short. The main reason is that the level of design, construction, and reinforcement was very low, for example, flood control standard was low, or there were obvious shortcomings in dam structures and construction quality, or the technology of operation and maintenance were limited. Since the 20th century, the design and construction of concrete gravity dams have been standardized and technologies in reinforcement are getting more and more advanced, which have led to longer life span of concrete dams. Life span of a dam not only depends on the quality of dam but on the environment and the needs of the society. At the same time, it also has close relationship with the reinforcement. Life span of concrete gravity dams can be divided into natural life span, environmental life span and economic life span. Natural life span mainly lies on the own conditions of dams. Because of structures, materials, earthquakes, floods or other reasons, some dams may become defective and need to be rebuilt, disused or removed, which can be considered to reach

their natural life span. For those which are breached due to natural reasons (excluding wars or terrors, etc.), it can also be considered to have reached their natural life span. For example, more than 3,000 dams in China and 1,000 dams in United States were breached and arrived at their natural life span. One gravity dam with height over 50 meters in Canada was reported that it had seriously problems after operating for more than 50 years, the cost of reinforcement may be higher than that of rebuilding a new one and the final decision is to rebuilt it in 2003. This can also be considered for the dam to reach its natural life span. When a dam need to be abolished because of reservoir silt or the needs of environmental protection or the changing of dam purpose, it can all be considered that the dam has already reach its environmental life span. After a period of operation, the security and functions of a dam are far below compared with a new dams, we can consider that the dam has already reach its economic life span. During the natural life span of a dam, there could be several economic life span cycles. Different dams have different economic life span. The normal service period of concrete structure is 50 years defined by some countries, but according to operation status of gravity dams around the world since 20th century, the economic life spans of gravity dams may be over 50 years. Besides, there are also some with service period less than 50 years. Considering the technology related and behavior of dam changing obviously with time, it has practical significance to carry out comprehensive evaluation and studies on dam safety at their economic life span and try to make the safety recover to the level of a new dam.

Another factor which affects reinforcement is the environmental life span of dams. Sometimes, the environmental life span may be obviously shorter than the natural life span and it will also have impact on the economic life span of the dam. For example, many rivers in the world are sediment-laden river and the sediment in reservoir will reduce the capacity of reservoir directly. Although measures have been taken to alleviate siltation to some extent, life span of such reservoirs sometimes is limited, even shorter than 100 years, which will affect the schemes of reinforcement. The reservoir abolishing will cause the ending of economic life span of dams.

According to statistics, among 87,076 reservoirs in China, there are about 37,800 with safety problems. To ensure the security, studies and reinforcement have been carried out in recent years. In the new schemes (2007–2009), the reinforcement of 6240 dams will be conducted. Some of them are very difficult to deal with. The main problems are as following,

(1) Flood control problems: Due to increasing of hydrology data and safety requirement, flood control standard of reservoirs can not meet with the new operation conditions and the discharge ability of reservoirs becomes insufficient.
(2) Seismic problems: According to "Seismic Parameter Distribution Map of China" (GB18306-2001) and current Specification, Safety considering seismic loading cases of many reservoirs can not meet with the current requirements.
(3) Stability of dams: Because of insufficient in dam section or cracks existed or joints openning, many dams have to be rehabilitated.
(4) The leakage and uplift problems.
(5) Crack and Aging problems.
(6) The metal structures and electrical equipment problems: Metal structures and electrical equipment are aging or seriously eroded that they can hardly operate normally, which have seriously affected the safety of reservoirs.
(7) Management facilities and observation equipments are not in good conditions.
(8) Reservoir silt and landslide.
(9) Freezing and sawing problems.
(10) others.

The large-scale construction and management on hydro projects have been conducted for 50 years in China and many effective methods and experiences on reinforcement and heightening have been accumulated. But problems of how to determine the economic life span, how to carry out the long-term safety evaluations of the dams should be further studied.

2 MAIN PROBLEMS AFTER NEW COMPREHENSIVE EVALUATION ON FENGMAN DAM

Fengman concrete gravity dam is situated at the main stream of the second Songhua River, 24 km to downstream Jilin City, Jilin Province. It is situated in severe cold area. Its mean annual temperature is 5.3°. The highest mean monthly temperature is 24.3° and the lowest is −19.7°. The maximum height of the dam for its original design was 90.5 m and the dam crest elevation is 266.5 m. Dam construction started in April 1937 and water impounding started in Nov. 1942. The project was completed and operated in Oct. 1953. By the end of reinforcement in 1996, the maximum dam height is 91.7 m and the dam crest elevation is 267.7 m. The dam crest is 1080 m long, divided into 60 dam sections, each of which is 18 m long. Arranged from the left to the right bank, the 9th to 19th dam sections of the dam are overflow dam sections, the 21st to 31st are intake sections for power generation. The upstream slope of dam section is 0.05 and downstream slope is 0.78. During the construction, the dam cross section was divided into A, B, C and D blocks by the longitudinal joints. The typical cross section after reinforcement is as shown in Fig. 1.

In order to guarantee the safety of old concrete dams over 50 years, it is really necessary to do the comprehensive evaluation based on studies on Fengman dam in China. Main problems and main achievements from evaluation up to now for the project are as following,

(1) Problems of stability related to seismic assessment

Original conclusion: With weak longitudinal joints and sub-longitudinal joints, the stress level of some position of Fengman dam is higher than the allowable value and the safety can not be guaranteed for seismic load and some parts of Fengman dam may be destroyed for used earthquake parameters. Based on results of original analyses, anchoring have to be installed and had been installed before 1997 from the top to foundation and the dam had been added to 91.7 m high (1.2 m higher than original dam) for improving the stability of the dam. With considering the reliability of anchoring and other issues, it is still a safety problem under seismic loading cases.

New conclusion: Considering new progresses made in past 20 years, seismic parameters have been comprehensively evaluated based on current standard. It is found that the acceleration coefficient can be decreased from 0.161 to 0.131 and not to 0.22 or even higher as early estimated. The new results have significant influence on future rehabilitation work.

(2) flood control

Original conclusion: The spillway consists of eleven 6m by 12 m orifice sluice ways. Discharge capacity of

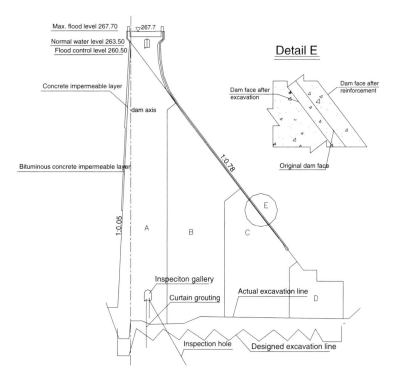

Figure 1. Typical cross section of dam.

1300 m³/s by the 10 turbines in the powerhouse is used for flood control before for the reservoir at El. 266.5 and El. 267.7.

New conclusion: The discharge capacity of the plant would not be allowed to use when the reservoir level at 267.7 to guarantee the safety. The flood control problems will be more critical for the current design standards compared with original ones in histories. It should be carefully evaluated to consider the possibility to use pre-discharge based on reliable hydrological monitoring and prediction system. Further measures to increase the capacity for flood control should be studied.

(3) Serious Leakage of the Dam and high uplift pressure

Original conclusion: With poor integrity and defects, such as crack and honeycomb, serious leakage from the dam body and joints occurred after impoundment, which affected the integrity and durability of dam. When the reservoir water level reached El. 255 m in 1950, leakage measured in the galleries reached 16,380 L/min and the wet area in the downstream surface was about 24947 m². After rehabilitation for many times, leakage today reduces to 39L/min totally and wet area in the downstream surface at El. 256.55 m is about 440 m² in 2004, most of which is located at spillway sections. While average uplift pressure coefficients for blocks of 8, 14, 22, 28, 35, 40, and 47 at

different position monitored in 1996 are 0.84, 0.48, 0.63, 0.16 and 0. The distances between test holes to the dam axis are 2.9 m, 6.5 m, 12 m, 39 m and 51.6 m respectively. Monitor data at block 15 in 2005 give similar results. Grouting measure could be reliable and has been done many times to decrease the leakage and uplift.

New conclusion: High uplift pressure is still a big problem to be solved in the future. Grouting measures can be used but not enough especially for decreasing the uplift in the body. Geomembrane installation under water can be a reasonable choice for further rehabilitation.

(4) Poor Quality of Concrete

Original conclusion: Strength of dam concrete of 90d in the original design should be 150 kg/cm², but actual strength of concrete was only 120 kg/cm², 90 kg/cm² or even 60 kg/cm² at different parts of dam. Aggregate and cement used were not in good quality. Water reducing admixture and air-entraining admixture agent were not used and there was no freezing resistant consideration for all concrete even though the dam locates at extremely cold area. Concrete mix proportion in construction had problems and there was no temperature control. Outer concrete of 0.6 m thick has been replaced in recent years during rehabilitation. While most part of concrete in upper and downstream side of dam have very low strength of about 50 kg/cm^2. Grouting measure can improve the situation.

New conclusion: Grouting measure is not enough. Drainage measure, thickening the dam body and etc. are necessary for further action.

(5) Longitudinal joints of Dam

Original conclusion: Fengman dam was divided into blocks by transverse joints, longitudinal joints, sub-transverse joints and sub-longitudinal joints. No special treatment was done to most of these joints. No shearing resistant structure and joint grouting was made in AB longitudinal joints from EL.220 to EL. 242 (Fig. 1). Although grouting was carried out during operation for many times, unbounded longitudinal joints of the dam body are still a problem. Model test, static and dynamic analyses made before show that the joints have big influence on stress distribution and stability. Strong anchoring measures should be carried out considering possible higher seismic load.

New conclusion: Strong anchoring measures are not so important as before.

(6) Freezing and Thawing Problem

Original conclusion: With poor quality of concrete and serious uplift pressure of the dam, some surface concrete was destroyed by freezing and thawing condition, especially for spillway sections. In 1986, the damaged concrete of the upstream surface above elevation 245 m and of the downstream surface from top to ground level were excavated 0.4 m and were covered by 1 m reinforced concrete with high strength and frost resistance. The dam crest was heightened by 1.2 m. For spillway, the downstream surface was replaced by new concrete with 1.5 m thickness and was fixed with lots of anchorage (3.5 m deep).

New conclusion: The freezing and thawing problem is still a big and challenging problems up to now and it will decrease the safety obviously. The original measures are not enough and concrete with thickness more than 4m should be put on the surface downstream to achieve similar safety standard as a new dam.

Main rehabilitation work finished before 1997 is following,

(1) Bituminous concrete lining with a thickness of 10 cm was placed on the upstream surface between El.245 m to El. 226 m before the flood season in 1990.
(2) Grouting to dam body and foundation was one of the main measures adopted for reducing the seepage and uplift pressure. Grouting was carried out in 37 dam sections in different years.
(3) Pre-stressed anchors were installed to improve the dam safety under earthquake loading cases. 378 pre-stressed cables in total with different loading grades were installed in No.7 to 49 dam sections, in which 361 cables were installed in dam body (excluding the test anchors) and 17 cables, in the dam foundation.

(4) Heightening of dam. According to the dam reinforcement design, the dam crest has been raised by 1.2 m to improve stability of dam and increase flood control ability.

The work above is reasonable but enough to achieve a new economic life span for the old dam.

3 SAFETY EVALUATION BASED ON NUMERICAL ANALYSIS BY FEM FOR FENGMAN DAM

To truthfully reflect the practical safety situation of the Fengman dam and achieve clear results compared with current dam, simulation analyses have been made with considering construction processes and longitudinal joints and construction joints. Whole course simulation analysis of BL47 from the construction period to the operation period is conducted. A three-dimensional FE model is built, which takes into account of all factors that will affect the stressing and deformation of the dam.

In simulation analysis process, the construction process is simulated according to construction data recorded, using measured data of air temperature and water level as boundary conditions. The calculation period is from October, 1941 to December 2005. In the concrete construction period, the minimum calculating step is 0.5d, while the maximum calculating step is 5d. In operation period, calculating step is 10d. There are 2,707 calculating steps in the whole simulation analysis process in total.

In whole course simulation, the adiabatic temperature rise of the concrete material in simulation analysis is determined according to practical mixing ratio of the concrete and the maximum internal temperature observed on site. The final elastic modulus, coefficient of linear expansion and coefficient of temperature conductivity are determined according to observed data of displacement and temperature by back analysis. Main parameters can be seen in Table 1.

According to the whole course simulation calculation, temperature field, stress field and displacement field of BL47 at any time can be obtained. Comparing the horizontal displacement at dam top obtained by simulation analysis with measured data (See Fig. 2), it can be found that these two are consistent with each other, which indicates that the simulation analysis can reflect the practical operating condition of the dam.

According to the envelope diagram of minimum temperature field during operation from 1990 to 2005 (Fig. 3), negative temperature may occur in concrete of dam top and downstream side during winter, which may cause frost thawing fracture due to sever seepage in dam concrete. The depth of negative temperature zone of downstream side is about 4.5 m. According to the contour diagram of σ_1 in winter (Fig. 4), tensile

Table 1. Value of material thermodynamic parameters.

Material	E (GPa)	Specific gravity (kg/m³)	Poisson's Ratio	Coefficient of temperature conductivity (m²/h)	Specific heat (kJ/kg·°C)	Coefficient of linear expansion (10⁻⁶/°C)
Bedrock	15	2750	0.21	0.00342	0.967	7
Concrete	11	2350	0.25	0.0043	0.978	7

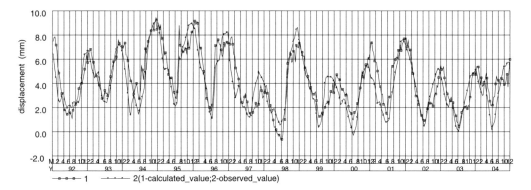

Figure 2. Calculated horizontal displacement compared with observed data at dam top of BL 47.

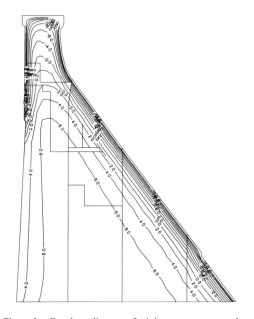

Figure 3. Envelope diagram of minimum temperature during operation (°C).

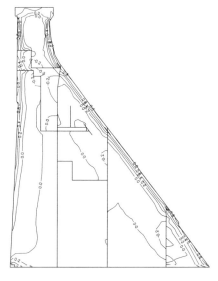

Figure 4. Contour diagram of σ_1 in winter (MPa).

stress with maximum value of 0.8 MPa is distributed in area of dam shell, in most part of which concrete has low tensile strength of about 0.5 MPa. The depth of tensile stress zone of downstream side is about 3.2 m. Furthermore, tensile stress of 0.2–0.8 MPa is distributed in dam heel with width of 6 m or so. Calculated results show that the value and distribution area of tensile stress in dam heel may decrease, while tensile stress distributed in concrete of dam center way cause longitudinal joints and construction joints to open.

Results of simulation analysis show that most part of longitudinal joints of Fengman dam have opened gradually even during the construction period. More

Table 2. Results of overloading analysis.

Items	Case 1	Case 2	Case 3
Overload coefficient	3.26	1.67	1.11–1.14

than 70% of longitudinal joints are opened up to now, which makes it impossible for the dam to bear load as a whole and has deteriorated the work behavior of the dam. According to calculation, safety factor of sliding resistance along dam base, when considering the affect of longitudinal joints and stress history, is 10–20% less than that evaluated by traditional methods, in which no such factors are taken into account. To some blocks of Fengman dam with faults passing through bedrock, potential risk of unstability under some work condition may exist.

Over loading analysis under different conditions has been carried out to evaluate the safety of Fengman dam compared with a similar new dam: (1) Case 1: overloading analysis on a newly built dam in nowadays which has the same dam section as Fengman dam and longitudinal joints are well dealt; (2) Case 2: overloading analysis on a dam with the same dam section and concrete quality as Fengman dam while longitudinal joints are well dealt; (3) Case 3: overloading analysis on Fengman dam under work conditions of nowadays. The results can be seen in Tab. 2. Obviously, poor concrete quality and affect of longitudinal joints are two important factors, which have led to the decrease of dam safety.

To increase the safety of Fengman dam, it is an option to add new concrete on the downstream face. When 4-meter-thick concrete layer is added, overload coefficient will increase to 2.33–2.47 even no reinforcement is done to longitudinal joints and safety factor of sliding resistance along dam base will increase more than 17%, which will meet the requirements of sliding resistance.

4 MAIN SUGGESTIONS FOR FUTURE REHABILITATION FOR FENGMAN DAM

(1) Adding new concrete with 4 to 6-meter-thickness on the downstream face

Thickening the dam along downstream face will increase the safety of the whole dam and also make negative temperature and tensile stress transfer to new high strength concrete, which will improve safety of the dam to a similar new one.

(2) Decreasing the leakage and uplift measures

Leakage, high uplift pressure, freeze and thaw situation are serious problems that can't be ignored. Fengman reservoir with more than 10 billion cubic meter's storage capacity is very difficult to empty. According to feasibility studies, several options have been put forward: (a) Install Geo-membrane on upstream surface with Carpi tech, which can be done partly under water; (b) Design and build a special removable coffer dam used to form a dry site so that it is possible to rebuild an upstream surface of dam; (c) Using New concrete core method to replace concrete of some distance to upstream face with new concrete part by part so as to form a concrete impervious wall; (d) others. Geo-membrane installation on the upstream face is the first choice combined with other measures up to now.

(3) New drainage system to decrease uplift pressure

It is clear that if new drainage system could be constructed, the uplift pressure especially for spillway sections will be decreased. Based on FEM analyses, to construct a drainage system and a new drainage tunnel 2 m under the downstream surfaces and about 40 m away from the dam axis could function well. This method has been used by Shuifen Gravity dam in China and demonstrated well for about 40 years operation located in similar condition.

(4) Further grouting on dam body

Grouting on dam body is still an option to improve the strength of concrete and control seepage. Experiment has been done on dam section 14 and 36, the results is not as good as estimated. Further studies will be made for achieving better results.

REFERENCES

(1) Jinsheng JIA, Yihui LU, etc. Comprehensive evaluation on Fengman dam, China Institute of Water Resources and Hydropower Research, June, 2007.
(2) Jinsheng JIA, etc. Solutions to Fengman dam rehabilitation, China Institute of Water Resources and Hydropower Research, June, 2007.

Tailor Made Concrete Structures – Walraven & Stoelhorst (eds)
© 2008 Taylor & Francis Group, London, ISBN 978-0-415-47535-8

The Rijksmuseum in Amsterdam

Antonio Ortiz

Cruz y Ortiz arquitectos, Sevilla, Spain

CURRICULUM VITAE

1948	Born in Seville.
1971	Degree in Architecture, School of Architecture, Madrid.
1971	Partnership with Antonio Cruz.

Academic experience

2004	Honorary Professor, **University of Seville.**
2002	Kenzo Tange Visiting Design Critic, Graduate School of Design, **Harvard University.**
2000–01	Visiting Professor, **University of Navarra, Pamplona.**
1998	Visiting Professor, Graduate School of Design, **Columbia University.**
1995–96	Visiting Professor, **University of Navarra, Pamplona.**
1994–95	Visiting Professor, Graduate School of Design, **Harvard University.**
1992–93	Visiting Professor, École Polytechnique Fédérale de **Lausanne (EPFL).**
1992	Visiting Professor, Department of Architecture, **Cornell University.**
1989–90	Visiting Professor, Graduate School of Design,**Harvard University.**
1987–89	Visiting Professor, Eidgenössische Technische Hochschule **(ETH), Zurich.**
1974–75	Professor, **School or Architecture, Seville.**

Tailor Made Concrete Structures – Walraven & Stoelhorst (eds)
© 2008 Taylor & Francis Group, London, ISBN 978-0-415-47535-8

Prefab, the American Way

Thomas d'Arcy on behalf of PCI

The European Standard for Prefabrication

Marco Menegotto
University of Rome, Italy

CURRICULUM VITAE

Family name:	Menegotto
First name:	Marco
Title:	Prof. Ing.
Date of birth:	31-01-1940
Nationality:	Italy
Present position:	Full Professor of Structural Engineering (since 1980), Università La Sapienza, Roma
Research:	Non linear analysis of concrete structures, Seismic engineering, Heritage structures, Prefabrication, Concrete material Coordinator of research programs of national importance Reviewer for the relevant journals and research centres
Design:	Consultant for design and rehabilitation of building and civil engineering structures

Participation in scientific bodies

Formerly

CNR	(National Research Council) Member of Commission on Structural Concrete
UNI	(Italian Standardization Agency) Member of Commission Structural Engineering
CEB	TG "Design by Testing": Member Commission "Structural Analysis": Member GTG "Prefabricated Structures": Reporter Commission "Buckling and Instability": Co-Chairman
RILEM	TC "Load-bearing Concrete Panels": Convener
EEC	Eurocode 2: Liaison member for Italy
IABSE	Permanent Commission "Concrete": Chairman
UNESCO	Consultant for assessment of stability of heritage constructions
FIP	Council: Vice-President, for Italy Commission "Prefabrication": Co-Chairman
fib	WG7.3: Member

Presently

AICAP	(Italian Structural Concrete Association) Vice President
ASSIRCCO	(Italian Association for Consolidation and Upgrade of Constructions) Vice President

CSLLPP	(High Council of Public Works of Italy) Expert member
CEN	TC250/SC2: Delegate TC229: Member of Steering and Coordination Group TC229 – TC250 Liaison officer
fib	Council: Delegate for Italy Commission Prefabrication: Chairman Special Activity Group for the New Model Code: Member
Honours	CTE (Italian Technical Building Society) Medal for Prefabrication
Sport	Rowing

Tailor Made Concrete Structures – Walraven & Stoelhorst (eds)
© 2008 Taylor & Francis Group, London, ISBN 978-0-415-47535-8

Fibres in Concrete, Research, Rules, Practice

E.H. Horst Falkner

ABSTRACT: Fibres in concrete is an innovative technic. Fibres have a steady increasing importance for structural and strengthening applications. Fibres are modifying the .... of concrete in a more ductile behaviour. The presentation gives an overall view about the progress of the last decade made research rules and application.

Concrete is the top material being used worldwide. Concrete enables any shape we want and meanwhile concrete strength reaches the one of steel.

We are pleased to have defined "strong and heavy fibres" such as reinforced steel bars or pre stressing steel in concrete structures.

Concrete with small fibres may be the *champion in material* for many future structures. Fibres in concrete made of steel, glass or polyurethane do not only improve the material in structural behaviour of concrete but are contributing for better serviceability behaviour, more robustness and durable structures.

The contributing will give an overall view to research, rules and application in practice of fibre reinforced concrete structures.

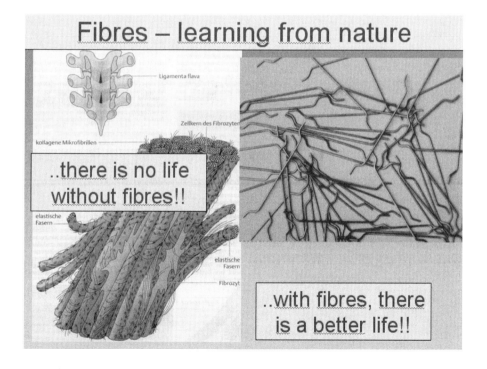

Tailor Made Concrete Structures – Walraven & Stoelhorst (eds)
© 2008 Taylor & Francis Group, London, ISBN 978-0-415-47535-8

Public Private Partnership, tool or trend

Andrea Echberg
Division Director Marquarie, London, UK

CURRICULUM VITAE

Background:

Andrea joined Macquarie in January 2006, prior to which she worked at ABN AMRO in the Infrastructure Capital team. Andrea has over ten years of project finance experience predominantly in the infrastructure sector. She started her career at London Underground's PFI advisory team working on some of the very first UK PFIs. Following that Andrea worked for two years in the Natural Resources project finance team at the Industrial Bank of Japan (now Mizuho) before joining ABN AMRO.

Experience:

Since joining Macquarie, Andrea has worked on a variety of infrastructure projects with a focus on developing PPP projects across Europe as a principal shareholder.

Andrea's key strengths include:

Structuring infrastructure projects from a sponsor perspective including contributing across the full spectrum of financial structuring, bidding strategy and leadership.

Negotiating commercial issues between the special purpose company and the public sector and construction and facilities management sub-contractors.

Andrea's recent project experience in infrastructure includes:

National Grid Transco RDNs, UK

€2.5 billion principal finance bid for the South of England regional gas distribution network including negotiations with NGT and its advisers, capital structuring, dbt structuring, equity, opeartor strategy and legal, technical, financial, regulatory and pensions and insurance due diligence.

Ministry of Finance Refurbishment PPP, the Netherlands

Director responsible for successful principal bid as 80% sponsor of the Safire consortium (ABN AMRO, Strukton, GTI, BurgersErgon and ISS Nederland) for the €120 million Ministry of Finance refurbishment PPP in the Hague.

Mestre Hospital PPP, Italy

Lead Arranger to the Astaldi led consortium for the €225 million Mestre Hospital PPP. This was the first PPP hospital in Italy with full FM risk transfer to the private sector. It was awarded European PPP of the Year award 2005 by Project Finance International.

N31 Road PPP, the Netherlands

Financial Adviser to the Combinatee Middelssee Consortium for the N31 road PPP in the Netherlands.

Armoured Vehicle Training Services PFI, UK

Financial Adviser to the preferred bidder, Landmark Training, for the Armoured Vehicle Training Services PFI including development of a pass/fail based payment mechanism.

Tubelines, UK

Lead Arranger to Tubelines for the London Underground PPP.

Tailor Made Concrete Structures – Walraven & Stoelhorst (eds)
© 2008 Taylor & Francis Group, London, ISBN 978-0-415-47535-8

The construction project of the Sagrada Família

Josep Gómez, Ramon Espel & Jordi Faulí

ABSTRACT: The Sagrada Família is Gaudí's unfinished work, to which he exclusively dedicated his last years of life. Even though he only got to build a small part of the total, he defined the rest through models and drawings.

Gaudí's design for the inside of the Temple was based on a new geometric architecture which made extensive use of ruled surfaces (paraboloids, hyperboloids), opening a new field which later architects have followed.

The following article aims at showing the construction complexity of these structures; especially the Temple naves with the vaults at 30, 45 and 60m heights will be discussed. This will show how the construction method is adapted to the construction needs according to the geometric shape, size, position, material repetitions of each vault.

1 FOREWORD

The Temple of the Sagrada Família is the last Cathedral to which the classic concept of an evolutive construction process, of over a century in duration, can be applied. Located in Barcelona's city centre, it is the most mature work of Antoni Gaudí, the architect *par excellence* of Catalan Modernism in the early twentieth century. Although Gaudí's initial philosophy is maintained, the construction process must change accordingly to the evolution of the time's building technology. Therefore, the evolution of its construction process closely mirrors that of the general building techniques of the epoch.

Gaudí's originality and innovative use of construction techniques are both present in the Sagrada Família.

2 THE SAGRADA FAMILIA DESCRIPTION

The Temple of the Sagrada Família, is a basilical, latin-cross plan building, with five longitudinal naves and three more naves forming the transept (Figure 1). It is surrounded by a rectangular cloister, with twelve perimetral towers, symbolizing the Apostles (95 to 115 meters high), crowning the two existing and another projected façades. Besides, six big towers will be built on top of the vaults. The central tower, measuring 170 m, represents Jesus Christ and is surrounded by the four 125 m tall Evangelists towers and the 120 m tower of the Virgin Mary, The tallest towers are situated

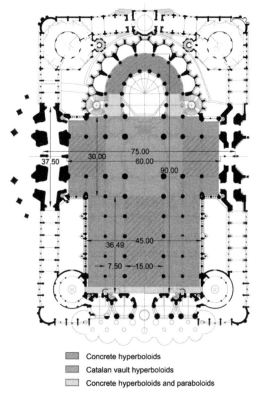

Concrete hyperboloids
Catalan vault hyperboloids
Concrete hyperboloids and paraboloids

Figure 1. Plan of the Temple of the Sagrada Família.

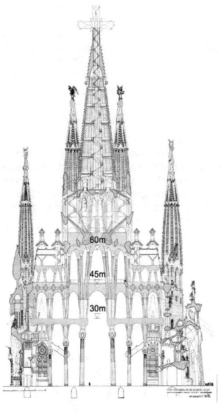

- ▢ Concrete hyperboloids
- ▢ Catalan vault hyperboloids
- ▢ Concrete hyperboloids and paraboloids

Figure 2. Section of the Temple through the transept, showing the location of the various types of vaults.

in the centre of the transept and on top of the apse, solving the maximum difficulty for the Renaissance architects, while the naves don't need buttresses, surpassing Gothic architecture by means of a static load equilibrium in its design.

One of the most important innovations of the Sagrada Família are its ruled surfaces (hyperboloids, paraboloids, helicoids, etc.), which will comprise the greater part of its elements (from windows to roofs).

Since 1910 Gaudí almost exclusively dedicated his efforts to the design of the Sagrada Família almost exclusively. In 1914 Gaudí designed a new space for the inside of the temple, with a new architecture based on ruled and other geometries (Figure 2). Gaudí's approach was to rationalise the structures by enhancing different aspects of Gothic architecture. These structures and branching internal columns, designed according to antifunicular load equilibriums,

Figure 3. Columns of the naves.

substitute the buttresses in their role of transmitting vertical loads to the foundations.

In the final solution, the columns will be double-turn helicoids, the roofing will be formed by paraboloids and the interior part of the vaults will consist on hyperboloids overlapping transition paraboloids. Their complex assembly was clearly defined in the 1/10 scale models and in their photographs. In 1936 his workshop was burnt down, but the models survived because they were made of plaster, thus resisting fire, allowing his followers to rebuild some of them.

Building aspects related to the use of concrete and precast concrete are also important. He decided that the vaults should form a monolithic form that would take the shape of a hyperstatic structure capable of withstanding a reduction in the number of some of its elements (arches or columns). This resulted in these ceilings being designed as hyperboloids constructed out of reinforced concrete, a material that can take on any shape and would enable Gaudí to design the vaults as unified structural elements. Figure 3 shows the tilting pillars of the main naves.

Gaudí designs and builds the pinnacles for the Nativity façade towers out of reinforced concrete. The four belfries of the towers of the Nativity Façade end in 24.6-metre-long spires built with a nucleus of reinforced concrete. Furthermore, the external pieces of the top 17 metres of the spires, which are covered in Murano glass, are crafted from prefabricated reinforced concrete pieces made in the onsite workshops.

It may seem that in these elements Gaudí used concrete in the same decorative fashion as in his earliest works. This, however, is not the case: the prefabricated pieces that top these spires are over 3 metres tall and are affixed to the building at a height of over 100 metres. This meant that they had to be built from a material that could resist the tension that could arise both during handling and in the final configuration. His response to this challenge was to use reinforced concrete. Over time, most of the architectural elements of the Sagrada Familia were found to be perfectly suited to being crafted in reinforced concrete.

3   EVOLUTION OF THE WORKS

After Gaudí's death, the construction process was continued. The fire that ravaged Gaudí's workshop in 1936 destroyed most of his plans and studies, but the architect Domènec Sugrañes continued with the construction despite moments of opposition from society. The construction process maintained the spirit of the earlier era through the substantial involvement by three of Gaudí's direct collaborators (Isidre Puig Boada, Francesc Quintana and Lluís Bonet Garí) and a search for innovation in the building process.

At this moment, the organisation of the progress of the project and of the construction of the Sagrada Família stands out from other current works, because of the characteristics and information of Gaudí's project and because the main responsibility, both for the project and the construction, belongs to the same organisation, The Construction Board of the Expiatory Temple of the Sagrada Família. This organisation is also the promoter. In the Temple, it has therefore been possible since the first planning phases, to have the collaboration of the different agents involved, so that the direct protagonists of future construction can be informed of project decisions and provide their experience. Thus, the preparation of the project and the construction are being done with the collaboration of the different specialists responsible for each phase of the process – the director architects, the architects responsible for the structures project, those in charge of the project and its construction and production. This relationship facilitates the quest for the best project solutions, taking into account faithfulness towards the original project, aesthetic and structural values and the construction method and process. On

the other hand, the project heads intervene when it is necessary to resolve problems during the production process or construction. All of this is possible because the departments in charge of preparing the project, of the production and the construction, work under the direction of the architect, Jordi Bonet, in the same office located under the main nave, where they prepare the documents necessary to ensure construction with quality, efficiency and in accordance with the project, and decide together on the programmes for coordinating it and establishing terms in order to achieve the objectives set by the Construction Board.

The structures project, commissioned to the office of the Facultative Direction architects Carles Buxadé, Joan Margarit and Josep Gómez requires constant communication with the technical office of the Temple in weekly meetings, and the fortnightly ones of the Facultative Management, in which the different proposals of the projects are assessed and improved from the structural point of view, both in the initial phases and in the construction. The structure projects, along with the systems to carry them out, are also issued and discussed.

Obtaining plaster models directly from computer drawings is fundamental in the entire process, because the architects can analyse a model in three dimensions soon after having drawn it and mounted it into the general models. They also have models in polystyrene expanded to natural scale, which are used to produce pieces in prefabricated concrete. Antoni Gaudí had a workshop of plaster modellers for producing 1:10 and 1:25 scale models and models for sculptures or large, prefabricated pieces for the pinnacles of the Nativity Facade. The workshop survives to this day and performs the same functions, with the aid of three-dimensional printers.

The construction of the 4,500 m2 of the interior of the Temple began with the main nave. The lower windows were begun in 1979 and in 1986, the columns foundation work was undertaken with concrete pilots sunk to a depth of 20 metres, with the subsequent construction, until the year 2000, of columns, windows and vaults for the aisle (in concrete to a height of 30 metres) and of the central nave (with a flat-brick vault at a height of 45 metres). The vaults of the transepts were then completed and work is currently being done on those of the centre of the transept at a height of 60 metres. In two years the space at a height of 75 metres will be complete, with a vault formed by a large hyperboloid; it will first be necessary to construct the vaults, at heights of 45 and 60 metres, which occupy a reduced circular crown around the centre of the apse.

The construction of each area or part of the interior of the naves involves new challenges of varying complexity, with different approaches, which must be resolved using the experience that has been obtained and by applying up-to-date building technology, in a

CONCRETE	ARCHITECTURAL ELEMENTS
MASS CONCRETE (fck 25 MPa)	• Fill walls and windows in the naves
REINFORCED CONCRETE (fck 45 MPa)	Structural elements with relative smaller loads • Columns of the naves • vaults of the central nave • Cloister • Glory façade • foundations
HIGH-STRENGTH CONCRETE (white: fck 80 MPa, grey: fck 60 MPa)	• Columns of the transept and crossing • Columns of the apse • Crossing windows • Columns of the Glory façade
SPRAY-APPLIED (fck 25 MPa)	• Vaulting in the central nave • Chorus
PREFABRICATED CONCRETE	• Upper part of the columns • Handrail of the chorus • Spiral staircase of the apse • Windows • Chapitels of the chorus.

process of improving building solutions and construction and construction and organisation methods.

4 BUILDING MATERIALS AND TECHNIQUES FOR THE TEMPLE NAVES

As we have seen, the building materials and systems used in the construction of the naves respond to the proposals in Gaudí's project – stone columns and windows, vaults of visible reinforced concrete and flat-brick vault (Catalan vault), stone roofs and Venetian glass mosaics. A structure of reinforced concrete, a material already used by Gaudí on the Nativity Façade, which he had also proposed for the vaults of the naves, begins in the foundations and finishes in the interior support structure for the roofs. The vaults at 30m are generally built of concrete. The hyperbolic vaults at 45 m. or higher up are being built using ceramic materials, as Gaudí had foreseen and as his collaborator and biographer Joan Bergós has written in his works.

In the early 1980s, and after the Passion Façade was completed, a geometric study of the Sagrada Familia was undertaken, starting with its columns and the internal vaulting of the naves. It was known that almost all the elements (columns, roofs, windows, domes, sacristies, etc.) are defined based on the intersection of ruled surfaces (hyperboloids, paraboloids, helicoids, etc.). After the analysis of the available models and drawings, the Sagrada Familia's technical staff was able to find mathematical equations for these figures. It was then necessary to find a way of representing them three-dimensionally that was quicker and more precise than Gaudí's plaster models.

Since 1991, a team led by Josep Gómez Serrano has used CADD-S5 software to draft new work on the church. This team is specialised in 3D modelling and is additionally prepared for working with the three-dimensional printer situated on the grounds of the Sagrada Familia. This allows those working in the modelling workshop to work with a great degree of precision, which significantly reduces the cost and time required to produce each new piece. These programs have become an indispensable tool in the ongoing construction of the Sagrada Familia.

Stone

The large stone elements of columns and windows have been produced either manually or with numerical control since 1989 using a disc-saw with two and a half axles and currently with five-axle machinery, with the participation of workshops with a long tradition, along with coordination and control from the Temple to ensure quality and coherence with the project as a whole. The lower columns have been built with the stones that Gaudí decided according to its resistance: sandstones from the Mountain of Montjuïc in Barcelona, granite, basalt and porphyry.

Prefabricated concrete pieces

Part of the columns, much of the windows and part of the elements of the railings and vaults have been constructed with prefabricated reinforced concrete pieces with stainless steel and with the texture of the four stones of the lower columns as covering and formwork of the structural reinforced concrete. Gaudí also used

Figure 4. Stone joint.

prefabricated pieces, for example in the Parc Güell, in the Colònia Güell Church, on the pinnacles of the Nativity Façade (concrete prefabricated pieces with Venetian glass and plaster moulds), etc.

Some of the 3 to 6-metre high prefabricated columns have been fabricated with incorporated steel reinforcement.

The models may be of plaster, as has been the case for the last ninety years in the Temple, or of polystyrene or polyurethanc with mechanisation. The moulds are made of these two materials and also of polyester and glass-fibre, depending on the number of pieces to be produced.

In order to reduce the fitting time, prefabricated pieces are mounted at the foot of the Temple, so that large elements may be positioned in their places. In the last months the windows of the apse at 45 m high have been constructed with that technique. We can see at the photograph the elevation of the reinforcement pieces and finally how the prefabricated concrete formwork is filled.

In order to fit certain pieces, a load positioner (hydraulic tripod) has been created and patented in the Temple, due to the need to move very delicate pieces with cranes and to position them with very smooth, precise movements in the work. It has a system of hydraulic pistons, which remote commands contract or extend, and the whole element hangs from the crane at the same time.

Columns made onsite using polyester formworks

The structure is first positioned, followed by the formwork into which the concrete will be poured. These

Figures 5,6. Windows of the apse (2008).

are columns of high-resistance concrete in the area of the transept and in the branchings of the columns of the apse.

The reinforced steel for the concrete is prefabricated in large elements in order to ensure it performs correctly and to advance the entire process by saving assembly time in the work. Large formwork pieces are also factory-produced, following the geometry of paraboloids and hyperboloids.

These columns, along with other columns and vaults currently being constructed, are made from high-resistance ($600\,kp/cm^2$–$800\,kp/cm^2$) concretes, produced in a central concrete works, located at the site itself. With these concretes, it is possible to construct the project with the sections decided by Gaudí, but with the demands of current regulations.

The foundations began to be poured for the naves in 1986 and the internal columns in 1991. These new structures contained a far greater amount of reinforcement than previously built parts of the building. This required a very workable concrete and finally a liquefier as an additive to guarantee the correct settling of the concrete. From that point on, liquefiers have been typically used in the concrete used in the building process at the Sagrada Familia (whether it is prepared offsite or mixed at the construction site).

Figures 7. Columns made outside with polyester formwork.

Figures 8. Spiral staircase of the glory Façade.

Spiral staircases

For the construction in concrete of the slabs of the spiral staircase, a system has been invented for constructing the many metres of stairs in the building. Using an integrated hydraulic system, a truck moves along metal rails and supports and helicoidally positions the polyester mould used to shape the base of the stair slab. Once the slab has been concreted, the system has a de-moulding mechanism to avoid the shape of the mould being altered. It is then moved upwards again along the rails, and continues in this fashion.

The spiral staircases of the Main Façade are constructed with a lower steel helicoid which also acts as a formwork and stone handrail.

Concrete vaults

The concrete vaults which represent hyperboloids, located at 30m over ground level, can be built using moulds as formwork, due to the fact that there are a great number of repeated elements (20). This solution is introduced in 1993, as an adaptation of ship-building techniques to architectural needs.

In that solution, the modules that have been repeated several times, have been modelled with plaster on a natural scale in the modellers' workshop in the Temple and polyester moulds have been produced above them, the divisions having previously been studied, in order to have a certain amount of comfort when erecting them and de-moulding the vault. Once positioned in their location onsite, they are first Gunite

High-strength concrete was first used in the construction of the columns of the transept, atop of which the Evangelist Towers and the central dome will rest (1998). Thus, the project requires structures that will be able to withstand the compression loads and that will also respect the shapes and diameters set forth by Gaudí. Increasing the strength of the concrete decreased the amount of steel needed in the dense reinforcements, facilitated the construction process and prevented the builders of the Sagrada Família from having to bulk up the diameter of the columns. The final choice was made for high performance white concrete to be poured in situ (fck = 80 MPa) rather than the prefabricated architectural concrete (fck = 35 MPa) used in the columns of the main nave.

The initial mix for white HSC for the Sagrada Família was proposed by the technical management of the project in collaboration with technicians from the Bettor company. The results of the test undertaken with different components were used to vary the mixes of the other concretes used. The aim of using this concrete is to meet the following technical performance goals:

- On-site workability (Slump test >25 cm)
- Low porosity upon drying, and good durability
- A minimum strength based on the needs of each structural element
- Furthermore, the white concrete should exhibit the appropriate colour

Figures 9.   Polyester moulds of the lateral naves.

mortar-sprayed. The formwork is then fitted and it is concreted.

In the vault modules that are not repeated, a system of stray moulds has been sought, with a single use adapted to the different shapes of the regular geometry used by Gaudí, using the most appropriate materials according to the shapes and measurements. Their production is rather manual and craftsman-like, but approaches the maximum possible mechanisation. These vaults are basically located where the apse and the nave meet, and in the inside of the Glory façade.

The capital or skylight hyperboloids have been produced with wooden ribs, cut in hyperboloid shapes and which on joining with others in a circular shape on different levels, form what will become the structure of the formwork, subsequently lined with 5 mm marine panelling, which enables it to be interturned and adapted as required.

Production of a hyperboloid with wood

The large-dimension paraboloids are made by producing an initial skeleton, based on metal bars welded together, which follow the directrices and generatrices of the element, subsequently lined with thin, MDF panelling, which is given a bath of water with acrylic so that it softens and adapts better to the desired shapes.

Apse Choir Gallery

The formworks of small, high curvature hyperbolic paraboloids are made using measured steel bars welded together side-by-side. The spaces that remain are then covered with a combination of epoxy resins and sand, a mixture which gives the bars the texture of a continuous surface.

Once all these elements have been produced, another, quite complex job begins to position them in the space and fit them together. They are then

Figures 10.   Hyperboloids produced with wooden ribs.

Figures 11.   Hyperbolic paraboloids.

underpinned, so that they can take the pressure of the concrete and finally the entire mould, which is exposed to the open air, is coated with paint in order to further protect it before it is poured.

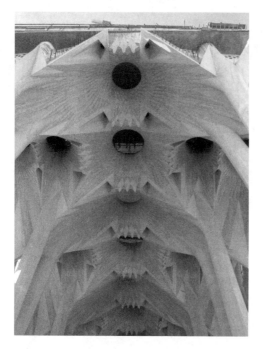

Figures 12.   Flat-brick vaults (Catalan vault) hyperbolic paraboloids.

Flat-brick vaults (Catalan vault)

This is a traditional Catalan construction technique for covering spaces and constructing stairs. Three layers of mortared tile or brick are superimposed to create a very resistant layered combination. At the nave of the Temple of the Sagrada Família, this solution was used for the first time in 1998, coinciding with the change of the main crane. They are basically located at the 45 m high vaults, both in the main nave and in the transept, at the 60 m vaults in the crossing, and in some zones of the 30 and 45 m vaults in the apse.

On the domes of the Temple, the first layer of tile follows the straight lines of the hyperboloids. In the spaces left by the rows of tiles, triangular elements are placed, decorated with green and gold glass, which represent the palm-leaves that Gaudí wanted on the domes. Construction and decoration follow the geometry and convert the flat-brick domes into an element that enriches the interior. The dome is constructed with a metal centering with the shape of a hyperboloid. This forms the guides that the bricklayer needs to correctly position the prefabricated tiles and triangles of the first layer correctly.

Figures 13.   Creuer voltes a 60 m d'alçada.

There are different contractors working on various areas, such as bricklaying, working with steel or stonemasonry, erecting of scaffolding or restoring the original building fabric. The directors and supervisors of the subcontracted companies or suppliers, as well as the managers of the woodwork, stonemasonry and maintenance workshops of the church building, all participate in the organisational process. The geometrical laws determined by Gaudí provide the common language for the collaboration of all the specialists, operatives and companies involved.

Large assembly platforms at different levels make the construction tasks easier and greatly assist in ensuring the safety of the workers and visitors, an aspect to which appropriate resources and efforts are designated. They are also especially useful when constructing the Temple's inclining columns and a great help when erecting new scaffolding, using surveying devices such as lasers to make them as accurate as possible.

Two years from now, the interior of the Sagrada Família will be completed, thanks to a coordinated effort from a coordinated team. Together they will have resolved the different challenges of its construction using a combination of traditional techniques of bricklayers and stonemasons, the qualified work of construction workshops and factories and the use of new computer and construction technologies. In this way, it will be possible to enjoy this extraordinary place that Gaudí bequeathed for the future.

Oral Presentations
Life cycle design

Tailor Made Concrete Structures – Walraven & Stoelhorst (eds)
© 2008 Taylor & Francis Group, London, ISBN 978-0-415-47535-8

Assessment of corrosion in reinforced concrete beams

D.I. Banic, D. Bjegovic & D. Tkalcic
Civil Engineering Institute of Croatia, Zagreb, Croatia

ABSTRACT: The effect of corrosion, due to chloride attack, on reinforcing steel and concrete was studied by performing a series of tests on thirty specimens. Beam specimens recommended by the RILEM/CEB FIP COMMITTEE were used. Specimens consist of two halves of a reinforced concrete beam, rotating around a hinge mechanism. Tidal and splash zone conditions are simulated by alternate wetting and drying cycles, using salt spray, which represents the typical seawater of the Adriatic Sea, in a large environmental chamber. The process of accelerated corrosion includes one cycle of wetting and drying per day. During the three-year experimental program, measurements of density of corrosion current, electrochemical potential and resistance were taken, using the linear polarization technique. Chloride penetration depth, $\times$ (mm) was calculated by means of known expressions. This paper describes measurement results and conclusions on corrosion penetration depth.

Tailor Made Concrete Structures – Walraven & Stoelhorst (eds)
© 2008 Taylor & Francis Group, London, ISBN 978-0-415-47535-8

Corrosion of reinforcement bars is not anymore inevitability

P. Guiraud
CIMBéton, Paris, France

F. Moulinier
Institut de Développement de l'Inox, Saint Herblain, France

ABSTRACT: Corrosion of the reinforcement bars is the main cause of the defacement of the works. It is principally due to the carbonation of the concrete and to the penetration of the chloride ions through the cover. Corrosion generates expenses for maintenance and repairs; it also has an effect on safety and serviceability and at last reduces significantly the life duration of the works. Faced with these environmental data three degrees of design strategies/approaches are possible:

– To build as cheap as possible and pay later for the repairs keeping in mind that the lifespan of the work may be very reduced
– To use processes in order to delay the occurrence of corrosion and, out of predictive models, manage the preventive repairs.
– To consider that corrosion is not inevitability, and that it can be durably avoided through the use of stainless steel rebars which do not corrode.

Stainless steel reinforcement bars are more expensive than carbon steel bars but they can be used partially just on the most exposed parts of the works, where corrosion may occur; this last solution consists in paying more initially but gain on through-life operational costs, and on durability. This approach which takes into account the global cost has been applied for many years to other industrial sectors and should be easily transposed to the construction sector in order to pass down to the future generations a durable heritage.

Tailor Made Concrete Structures – Walraven & Stoelhorst (eds)
© 2008 Taylor & Francis Group, London, ISBN 978-0-415-47535-8

New aspects in durability bridge design

M. Empelmann, V. Henke, G. Heumann & M. Wichers
Institute for Building Materials, Concrete Construction and Fire Protection (iBMB), Technical University of Braunschweig, Braunschweig, Germany

ABSTRACT: The present paper deals with design aspects of the life-cycle optimization for reinforced and prestressed concrete bridges. At the beginning the special requirements regarding the life-cycle of bridges, in comparison to normal concrete structures, will be shown. The minimum life-cycle of the overall bridge construction will then be compared with the life-cycle of single bridge elements. By linking the expenditure for maintenance and repair with the life-cycle of single bridge elements, weak points can be identified. Based on these results effective approaches to increase the life-cycle of the total structure will then be presented. By means of a typical example, possibilities able to quantify the effectiveness of optimization measures will be shown.

Tailor Made Concrete Structures – Walraven & Stoelhorst (eds)
© 2008 Taylor & Francis Group, London, ISBN 978-0-415-47535-8

Carbon dioxide as a stimulus for life cycle thinking in cement and carbon neutral concrete building

P.A. Lanser & A.M. Burger
Cement&BetonCentrum, 's-Hertogenbosch, The Netherlands

ABSTRACT: Sustainability covers all social, economic and environmental aspects of a product all over its life cycle. It requires both life cycle thinking and multidisciplinary skills. In the cement and concrete industry the key word 'sustainability' is mostly linked to the depletion of natural resources, the use of energy, the reuse of secondary materials and to emissions. Yet there are other faces of cement and concrete. Cement factories are often a place where alternative raw materials and non-fossil fuels are used. And concrete helps to save energy in the use phase. It is time to make a new over-all balance!

Tailor Made Concrete Structures – Walraven & Stoelhorst (eds)
© 2008 Taylor & Francis Group, London, ISBN 978-0-415-47535-8

Service life design of concrete structures by numerical modelling of chloride ingress

M.M.R. Boutz & G. van der Wegen
INTRON, Sittard, The Netherlands

P.E. Roelfstra & R. Haverkort
Femmasse, Sittard, The Netherlands

ABSTRACT: A numerical model is presented to simulate chloride ion ingress in concrete structures exposed to variable climatological conditions. The transport mechanisms of (free) chloride ions are both diffusion through and convection by pore water. This requires an advanced model for moisture transport in both the saturated and the non-saturated area. In case that the surface of the concrete structure is in contact with liquid water, the rate of the front of the saturated area is controlled by the balance between sorption and diffusion. The chloride binding isotherm is described by a linear or Langmuir type relation. The salient practical features of the model are: multi-layer, replacement of layers, imposing a chloride profile as initial situation. The model allows to predict the effect of maintenance actions on the service life of structural concrete, as illustrated for the King Fahd Causeway in the Persian Gulf.

Tailor Made Concrete Structures – Walraven & Stoelhorst (eds)
© 2008 Taylor & Francis Group, London, ISBN 978-0-415-47535-8

Influence of curing on the pore structure of concrete

W.J. Bouwmeester – van den Bos
Faculty of Civil Engineering and Geosciences, Delft University of Technology, Delft, The Netherlands
BAM Infraconsult, Gouda, The Netherlands

E. Schlangen
Faculty of Civil Engineering and Geosciences, Delft University of Technology, Delft, The Netherlands

ABSTRACT: Most concrete structures are designed to last for at least a hundred years or more. During this lifetime the structure is exposed to several environmental influences. Whether a concrete structure can resist these environmental influences depends, among other things, on the ingress rate of liquids and gasses. The rate of ingress has a direct relation with the pore structure and its connectivity, inside the concrete. The pore structure and its connectivity are influenced by several factors during design and construction. An important factor during the construction phase is the curing of concrete. To achieve the requested durable structure the concrete has to be cured. To perform this curing several methods are applicable: liquid membrane, plastic film, fresh water etc.

To investigate the effect of each curing method experiments are performed with two types of cement (CEM I and CEM III/B). The experiments consist of water penetration and rapid chloride migration tests. In the paper the results of the experiments are presented. Based on the results can be stated that concrete made with CEM III/B is more sensitive for curing than CEM I and it seems that water curing for concrete made with CEM I is less effective than for concrete made with CEM III/B.

Tailor Made Concrete Structures – Walraven & Stoelhorst (eds)
© 2008 Taylor & Francis Group, London, ISBN 978-0-415-47535-8

Life cycle management of infrastructures: Integration of disciplines as key success-factor

Aad van der Horst
Faculty of Civil Engineering, Delft University of Technology, Delft, The Netherlands

ABSTRACT: This paper addresses the integral approach in the development of infrastructural schemes, with a strong focus on the interaction between design and construction aspects. Key aspects addressed are Constructability, Reliability and Economy. An interaction scheme, to develop alternative concepts, is presented as well as a selection tool to make choices between alternatives developed. Examples are presented to support the interactions presented.

Tailor Made Concrete Structures – Walraven & Stoelhorst (eds)
© 2008 Taylor & Francis Group, London, ISBN 978-0-415-47535-8

Residual service-life of concrete façade structures with reinforcement in carbonated concrete in Nordic climate

J.S. Mattila & M.J. Pentti
Tampere University Of Technology, Tampere, Finland

ABSTRACT: The repair strategy is usually decided after a condition investigation. It reveals the critical cover depth i.e. the zone where reinforcement is under the risk of corrosion. If the corrosion is not widespread, the reinforcement with insufficient cover is usually chiseled out and patched. The rate of corrosion is usually not taken into account. However, the average rate of corrosion and the resulting residual service-life may vary within a wide range. In this study, corrosion rates of steel reinforcement in carbonated concrete have been monitored under the real climatic exposure in southern Finland. This revealed that the average rates of corrosion vary a lot between different types of structures. The rough residual service-lives resulting from that were found to vary from 5 to more than 50 years depending on the moisture exposure. So, in certain conditions, long service-lives can be achieved even though the reinforcement lies in carbonated concrete.

Tailor Made Concrete Structures – Walraven & Stoelhorst (eds)
© 2008 Taylor & Francis Group, London, ISBN 978-0-415-47535-8

The role of Controlled Permeability Formwork in life cycle design

Philip McKenna
Halcrow Group Ltd, Glasgow, Scotland

Chirag Baxi
Gujarat Narmada Valley Fertilizers Company Ltd, Bharuch, Gujarat, India

ABSTRACT: In recent years the number of reinforced concrete structures experiencing premature deterioration has grown considerably. It is worth noting that in many cases the structures in question were constructed less than 20 years ago. However, most were designed with an anticipated design life of approximately 100 years. This has forced governments to make provisions in routine maintenance budgets for structural rehabilitation, a process that is both expensive and also disruptive to the travelling public.

This paper shall compare concrete cast against conventional Impermeable Formwork (IMF) and demonstrate how a Controlled Permeability Formwork (CPF) liner can reduce initial construction costs, whilst at the same time achieving durability, through the natural enhancement of the near surface cover.

Tailor Made Concrete Structures – Walraven & Stoelhorst (eds)
© 2008 Taylor & Francis Group, London, ISBN 978-0-415-47535-8

Designed performance sustainable concrete

Boudewijn M. Piscaer
UNIVERDE Agencies sarl, France

Svein W. Danielsen
SINTEF Building and Infrastructure, Norway

ABSTRACT: The term "Designed performance sustainable concrete" is introduced as an alternative or a supplement to traditional Prescription concrete. The aim is to obtain more environmentally friendly materials and structures by selecting aggregate resources and binders, and doing structural design and mix design with a scope of reducing the total environmental impact at all stages of the process.

Tailor-made concrete structures – Case studies from projects worldwide

C.K. Edvardsen
COWI A.S., Kongens Lyngby, Denmark

ABSTRACT: There is a rapidly growing international demand for long-term well-performing concrete structures without premature need for maintenance and repairs. Major structures like bridges and tunnels are expected to have a long service life in the order of 100, 120 or even more years. Past decades have shown that the classical procedures for durability of reinforced concrete structures have often failed to provide reliable long-term performance in aggressive environments. Within Europe this awareness has led to the development of new service life design approaches to provide necessary and valuable tools to satisfy present day design needs. These new service life design tools have been implemented in **fib** bulletin 34 *Model Code for Service Life Design* and will be part of the new **fib** bulletin 34 Model Code for Service Life. The paper presents one of today's most modern durability design methodologies exemplified by some case studies.

Tailor Made Concrete Structures – Walraven & Stoelhorst (eds)
© 2008 Taylor & Francis Group, London, ISBN 978-0-415-47535-8

Rebuilding Le Corbusier's World Exhibition Pavilion; The Poème Electronique in Brussels, 1958

R. Nijsse
ABT, University of Delft, Delft, The Netherlands

ABSTRACT: The mythical building Le Corbusier has designed for the 1958 World Exhibition in Brussels was the first building to combine a light and sound show with Architecture. It was demolished after the Exhibition but the wish to rebuild this Pavilion is vivid. In 2006 a study was made how the Pavilion was built and how it could be done more efficient in our times.

Tailor Made Concrete Structures – Walraven & Stoelhorst (eds)
© 2008 Taylor & Francis Group, London, ISBN 978-0-415-47535-8

Why historic concrete buildings need holistic surveys

H.A. Heinemann

Chair of Building Conservation, Faculty of Architecture, Technical University of Delft, Delft, The Netherlands

ABSTRACT: With the increasing listing of 20th century buildings as monuments, it is necessary to find suitable conservation methods as these monuments differ from the traditional heritage. For concrete, one of the major 20th century building materials, a tailored conservation approach is needed urgently. At present, no specific concrete conservation approach exists, and concrete repair approaches are consulted. Looking at practice, cultural-historical values are often lost due to interventions. One reason is the insufficient awareness of possible values of the original concrete, which are thus not respected in the process. In this paper, the diverging aims of repair and conservation, which are seldom considered, are discussed. A closer look at backgrounds of repair and conservation explains why standard surveys endanger monumental values. To improve concrete conservation, a holistic survey integrating the technical condition and the monumental values is essential in order to obtain a sound basis for the conservation process.

Design strategies for the future

Tailor Made Concrete Structures – Walraven & Stoelhorst (eds)
© 2008 Taylor & Francis Group, London, ISBN 978-0-415-47535-8

Expansion joints with low noise emission

T. Spuler, G. Moor & C. O'Suilleabhain
Mageba SA, Bülach, Switzerland

ABSTRACT: Expansion joints can be a significant source of noise, if not carefully selected and correctly installed. New technologies to address this issue are therefore likely to increase in demand. This paper explores the different types of "quiet" expansion joints available, and new technologies which can transform an otherwise noisy expansion joint into a much quieter one.

Tailor Made Concrete Structures – Walraven & Stoelhorst (eds)
© 2008 Taylor & Francis Group, London, ISBN 978-0-415-47535-8

Overview of PCI'S research and development program

C. Douglas Sutton
Civil Engineering, Purdue University, West Lafayette, Indiana, USA

Paul Johal
Research & Development, Precast/Prestressed Concrete Institute, Chicago, Illinois, USA

ABSTRACT: Research is an integral part of Precast/Prestressed Concrete Institute's (PCI) Strategic Plan. Through continuous investment, the precast concrete industry continues to refine solutions and is actively pursuing new and emerging technologies. To meet the short-term and long-term needs of the industry, several different avenues have been used to sponsor research. These include research fellowships, specially funded research projects, research and development committee projects, and cooperative research projects.

Most of the major research programs are being carried out in collaboration with various universities, research organizations, governmental agencies and industry contributors. PCI funded research has grown substantially over the years. It has played a significant role in influencing code improvements, and it has provided a measurable impact on precast prestressed concrete industry practice.

Tailor Made Concrete Structures – Walraven & Stoelhorst (eds)
© *2008 Taylor & Francis Group, London, ISBN 978-0-415-47535-8*

Research on volume change movement and forces in precast concrete buildings

G. Klein & R. Lindenberg
Wiss, Janney, Elstner Associates, Inc., Northbrook, Illinois, USA

ABSTRACT: Volume change effects are the combined result of creep, shrinkage and temperature strains. Although a frequent cause of distress, volume change is often ignored in the design of precast concrete buildings. Design for volume change effects is complicated by the unknown influence of flexible connections and the extreme variability of concrete strain and the resulting forces. The objectives of this research are to 1) develop a better understanding of volume change effects based on measured performance of precast structures, and 2) recommend revised design procedures that reflect this understanding and account for the influence of flexible connections. The research program includes field monitoring of movements and strains of four newly constructed parking structures, development of computer models that accurately predict volume change behavior, and reevaluation of design procedures.

Tailor Made Concrete Structures – Walraven & Stoelhorst (eds)
© 2008 Taylor & Francis Group, London, ISBN 978-0-415-47535-8

The PCI headed stud anchorage research program: Scope and highlights of findings

Neal S. Anderson
Concrete Reinforcing Steel Institute, Schaumburg, Illinois, USA

Donald F. Meinheit
Wiss, Janney, Elstner Associates, Inc., Chicago, Illinois, USA

ABSTRACT: The construction industry has relied on the headed stud anchor to provide connection between concrete and multiple other structural elements for many years. In the past, the design engineer followed numerous accepted rational methods of designing for allowable strength and serviceability in accordance with accepted standard procedures.

Recent changes in the requirements in ACI 318, as well as new developments from testing completed by PCI, have started to impact the design engineer's practice in designing anchorages to concrete. Failure to stay current with new design procedures can result in liberal design solutions.

The historical overview discusses the chronology of design methods based on various boundary conditions and load applications. The standard for designing headed stud anchor connections has been impacted by new testing results compiled by PCI. This article is an overview of the current state-of-the-art approach used by precast structural designers for headed stud anchor design.

Tailor Made Concrete Structures – Walraven & Stoelhorst (eds)
© *2008 Taylor & Francis Group, London, ISBN 978-0-415-47535-8*

Development of a seismic design methodology for precast concrete floor diaphragms

R.B. Fleischman
Department of Civil Engineering and Engineering Mechanics, University of Arizona, Tucson, Arizona, USA

C.J. Naito
Department of Civil and Environmental Engineering, Lehigh University, Bethlehem, Pennsylvania, USA

J. Restrepo
University of California San Diego, La Jolla, California, USA

ABSTRACT: A multi-university research project is being performed to develop a comprehensive seismic design methodology for precast concrete floor diaphragms in the United States. The effort, funded jointly by the Prestressed/Precast Concrete Institute, the National Science Foundation, and the Charles Pankow Foundation, involves an integrated analytical/experimental research approach and strong industry oversight. Full-scale experiments of isolated diaphragm details were used to build analytical models of diaphragms for use in nonlinear static ("pushover") analyses and nonlinear transient dynamic analyses. New details have been developed and are being tested through hybrid (adaptive) experiments of precast panels integrated with computer dynamic analyses. Diaphragm force amplification factors, capacity design rules, and diaphragm detail classifications are being developed for the emerging seismic design methodology. A half-scale shake table test of a three-story diaphragm sensitive structure (one floor of untopped double tee, topped double tee and topped hollow core) will be performed to demonstrate the design methodology.

Tailor Made Concrete Structures – Walraven & Stoelhorst (eds)
© 2008 Taylor & Francis Group, London, ISBN 978-0-415-47535-8

Development of a rational design methodology for precast L-shaped spandrel beams

G. Lucier, C. Walter, S. Rizkalla & P. Zia
North Carolina State University, Raleigh, NC, USA

G. Klein
Wiss, Janney, Elstner Associates, Inc., Northbrook, IL, USA

D. Logan
Stresscon Corporation, Colorado Springs, CO, USA

ABSTRACT: Precast concrete spandrel beams are commonly used in parking structures to transfer vertical loads from deck members to columns. These beams typically have slender, unsymmetrical cross-sections and are often subjected to heavy, eccentric loading. These factors produce a complex internal structural mechanism including significant out of plane behavior. Traditionally, slender spandrel beams have been reinforced using the torsion and shear provisions of ACI-318. These provisions assure torsional strength by requiring heavy closed reinforcement and well-distributed longitudinal steel to provide torsional resistance after face shell spalling. The need for such complex and expensive reinforcement in slender spandrel beams is questionable. From as early as 1961, extensive field observations and limited full-scale testing have documented a lack of face shell spalling and spiral cracking in slender spandrel beams. Rather, out of plane bending appears to dominate slender spandrel behavior. More recently, extensive full-scale testing has confirmed that the classically assumed mode of torsional distress is not realized in precast slender spandrels. This paper provides background information and argument for simplifying the reinforcement detailing requirements for slender spandrel beams. The paper presents an overview and selected results to date of research currently in progress to develop alternative, more efficient reinforcing schemes and a rational design approach.

Underground structures

Tailor Made Concrete Structures – Walraven & Stoelhorst (eds)
© 2008 Taylor & Francis Group, London, ISBN 978-0-415-47535-8

Real opportunities for ultra high strength concrete in tunneling

T.W. Groeneweg
Movares, Utrecht, The Netherlands

C.B.M. Blom
Delft University of Technology, Public Works Rotterdam, The Netherlands

J.C. Walraven
Delft University of Technology, Delft, The Netherlands

ABSTRACT: From history of soft soil shield driven tunnels it is known that the required thickness of concrete tunnel linings is proportionally related to the tunnel diameter. The thickness should be approximately 1/20th of the diameter (D/20). It turns out that several structural mechanisms are dominating the required thickness at different boundary conditions. The structural mechanisms have been and still are intensively discussed.

The use of steel fibre reinforced concrete resulted in some very slender structures for bridges and roofs already. It turns out that the application of (ultra) high strength concrete is a real opportunity for tunnel linings. The application of higher strength concrete in tunnelling can result in cost savings in two ways: either by reducing the thickness of the elements or by reducing the amount of traditional reinforcement.

This paper presents the results of an extensive research to the influence of structural mechanisms on the lining thickness and discusses the opportunities of ultra high strength concrete.

Tailor Made Concrete Structures – Walraven & Stoelhorst (eds)
© 2008 Taylor & Francis Group, London, ISBN 978-0-415-47535-8

North/South Line Amsterdam, underground station CS on Station Island – Complex building techniques on an artificial island

R.M. van der Ploeg, J. Dorreman & J.C.W.M. de Wit
Adviesbureau Noord/Zuidlijn/Royal Haskoning

ABSTRACT: Since late 2002 the new metro line in Amsterdam has been under construction. The North/South Metro Line in Amsterdam, containing eight stations and measuring 9-km in length, will connect the northern and southern suburbs with the city centre (figure 1). As a requirement, the city centre's design solutions were specifically tailored to protect the historic structures and limit disruptions. The design of the underground station at Amsterdam's Central Station (CS) was largely determined by local environmental constraints. The underground station is being built in front of, behind and also underneath the historic central railway station; damage to the listed station building and obstruction of the existing traffic and passengers flows are not acceptable. An additional complication in the design process was the soft and highly variable subsoil. This article discusses the background of the design of the CS underground station.

Tailor Made Concrete Structures – Walraven & Stoelhorst (eds)
© *2008 Taylor & Francis Group, London, ISBN 978-0-415-47535-8*

Tunnels of improved seismic behaviour

S. Pompeu Santos
Civil Engineer, Lisbon, Portugal

ABSTRACT: The paper presents an innovative and cost-effective solution for the construction of tunnels of the roadway and railway types, executed with TBMs ("tunnel boring machines") when the referred tunnels are executed in soft soils (e.g., alluviums), in seismic areas, designated "TISB" concept. This concept can also be used for the strengthening of existing tunnels, using the existing tunnel as a formwork for the execution of the interior strengthening. In this paper a description of the solution is presented, as well as its application to a specific case.

Tailor Made Concrete Structures – Walraven & Stoelhorst (eds)
© 2008 Taylor & Francis Group, London, ISBN 978-0-415-47535-8

Behaviour of 'segmental/meter panels' for basements and subways

D.K. Kanhere
STUP Consultants Private Limited, Navi Mumbai, India

V.T. Ganpule
V. T. Ganpule & Associates, Mumbai, India

ABSTRACT: Growing need of basements for vehicles storage and services in city of Mumbai is problematic. Proximity buildings, services, narrow lanes, traffic and crowds preclude open excavations or diaphragm walls. Building subways under crowded roads with utilities and closure is not possible, led to using segmental/meter panel construction. These are panels of small width designed as cantilever sheet piles adequately anchored in sub-stratum to develop necessary passive resistance and balancing couple. When constructed in succession, full length of the wall can be constructed. This is a fast, flexible and cheap engineering solution to the problems encountered in crowded localities. A number of such structures have been constructed. Some panels were instrumented for verification of in-situ behaviour. Points of interest concerning stress patterns and distribution were monitored. A comparative study of the observed readings of the instrumented meter panels and the results of theoretical analysis and design principles are presented.

Tailor Made Concrete Structures – Walraven & Stoelhorst (eds)
© 2008 Taylor & Francis Group, London, ISBN 978-0-415-47535-8

The *New* Rijksmuseum – The art of going underground

D.D.J. Grimmelius & A.M. de Roo
ARCADIS, The Hague, The Netherlands

ABSTRACT: The Rijksmuseum in Amsterdam is one of the most important 19th Century monuments in the Netherlands. Currently the museum is redesigned by Cruz y Ortiz Arquitectos to meet modern demands. The main part of the renovation is the construction of a large central square of $3000\,m^2$. This square will serve as the new entrance to the museum. It is located in the existing courtyards and underneath the public passage running through the museum. To construct this square and its underlying facility rooms, a concrete basement with a depth of 6 meter is constructed very close to the existing wooden pile foundations. The relaxation of the soil due to the excavation and the construction method of the building pit have a large influence on the remaining bearing capacity of the existing piles and the settlements that are introduced in the building. Furthermore, a circular installation basement will be constructed around the entire museum with an approximate length of 400 meters. One of the main design problems is the integration of the new installations (especially mechanical ventilations) in the existing masonry structure. By providing this installation ring (internal dimensions 2.0 by 3.5 m) the museum can be entered at strategic places so that relatively small ducts can be used. A complicating factor for the design is the combination of a high groundwater level and the presence of water-carrying layers in the top soil. This has large consequences for the construction methods.

Tailor Made Concrete Structures – Walraven & Stoelhorst (eds)
© 2008 Taylor & Francis Group, London, ISBN 978-0-415-47535-8

Corrosion monitoring for underground and submerged concrete structures – examples and interpretation issues

R.B. Polder, W.H.A. Peelen & G. Leegwater
TNO Built Environment and Geosciences, Delft, The Netherlands

ABSTRACT: Since about 1980 Corrosion Monitoring Systems have been used in many concrete structures in aggressive environment worldwide. While these systems work properly in aboveground environment, some questions have arisen for submerged conditions, e.g. the outer sides of tunnels, piers in seawater or foundations in wet soil. One question concerns macro-cell formation between reinforcement in submerged concrete and in nearby aerated concrete, which might lead to severe corrosion for certain types or configurations of structures ("hollow leg"). In addition, in some cases, unexpected monitoring signals have been measured in submerged structures. The interpretation of electrochemical methods for monitoring the corrosion activity is not straightforward and new criteria have to be developed. This paper reports on an example of corrosion monitoring of an underground structure, the Green Heart Tunnel in The Netherlands. New criteria for interpretation of underground corrosion monitoring are proposed.

Tailor Made Concrete Structures – Walraven & Stoelhorst (eds)
© 2008 Taylor & Francis Group, London, ISBN 978-0-415-47535-8

Designing cast-in-situ FRC tunnel linings

B. Chiaia, A.P. Fantilli & P. Vallini
Politecnico di Torino, Torino, Italy

ABSTRACT: New procedures to design cast-in-situ steel fiber reinforced concrete (SFRC) tunnel linings are briefly presented in this paper. The ductile behavior at ultimate limit stage of such cement-based structures is ensured by a suitable amount of steel fibers and ordinary steel bars. The capability of SFRC to carry tensile stresses, also in the case of wide cracks, allows designers to reduce the minimum area of ordinary steel reinforcement, generally computed in compliance with American or European code requirements. In the serviceability stage, to evaluate crack widths more accurately, a suitable block model is introduced. This model is able to take into account the bridging effect of fibers, as well as the bond slip phenomenon between steel bars and SFRC in tension. Through the combinations of steel fibers and traditional reinforcing bars, it is possible to reduce the global amount of reinforcement in the structure, and contemporarily to increase the speed of construction. Consequently, the global cost of tunneling is reduced as well, particularly in massive structures. The proposed approach has been successfully applied to the design of two different tunnel linings in Italy.

Tailor Made Concrete Structures – Walraven & Stoelhorst (eds)
© 2008 Taylor & Francis Group, London, ISBN 978-0-415-47535-8

Concrete tunnel segments with combined traditional and fiber reinforcement

G. Tiberti & G.A. Plizzari
University of Brescia, Brescia, Italy

J.C. Walraven
Delft University of Technology, Delft, The Netherlands

C.B.M. Blom
Delft University of Technology and Public Works Rotterdam, The Netherlands

ABSTRACT: The paper deals with the concrete lining behaviour at Serviceability Limit State (SLS) in order to evaluate the advantages that result from an optimized reinforcement based on the combination of rebars and fibers with respect to the crack behaviour of segmental lining. For Serviceability Limit State, an analytical model was developed to describe the tension stiffening of a concrete element reinforced with traditional rebars and fibers. A parametric study was carried out to better understand the behaviour of segmental lining with different tunnel depth projections. It is shown that fibers can substitute part of conventional reinforcement and, as additional benefit, significantly improve cracking behaviour of the segment.

Tailor Made Concrete Structures – Walraven & Stoelhorst (eds)
© 2008 Taylor & Francis Group, London, ISBN 978-0-415-47535-8

Fire resistance of concrete tunnel linings

Jan L. Vitek

Metrostav a.s, and Czech Technical University, Prague, Czech Republic

ABSTRACT: Final tunnel linings are often designed made of only plain concrete. If such lining is subjected to fire, there is a substantial danger that a spalling would reduce the thickness of the lining more significantly than in the case of reinforced lining and also that the danger of spalling would threaten the rescue activities. The two series of tests when large scale concrete specimens were subjected to the fire according to the RWS temperature loading curve for a period of 180 minutes are described. The performance of concrete elements without and with polypropylene fibres was observed. After the tests the compression strength of concrete was measured on the cores drilled from the specimens.

Monitoring and inspection

Tailor Made Concrete Structures – Walraven & Stoelhorst (eds)
© 2008 Taylor & Francis Group, London, ISBN 978-0-415-47535-8

Segmental bridge behavior during bridge testing

M. Zupcic, D. Banic, D. Tkalcic & Z. Peric
Civil Engineering Institute of Croatia, Zagreb, Croatia

ABSTRACT: The majority of bridges in Croatia was built with prefabricated girders. A significant number of them has standard T and I cross section with continuous slab. These girders have considerable weight and there are many problems in transportation and placement in their position. Based on the preliminary analysis, it was decided that the so called BULB Tee cross section would be a convenient choice. Idealized cross section was designed according to geometrical characteristics and traffic load. The Civil Engineering Institute of Croatia followed their progress during construction and tested the bridges before they were opened to traffic. Both static and dynamic testing was performed. This paper discusses differences between bridge structures, results of investigations (deflections, transverse load distribution, dynamic characteristics such as frequency, damping, etc.). As a result of in-situ testing of these bridges, some recommendations for designing these new structures are given for bridge designers.

Tailor Made Concrete Structures – Walraven & Stoelhorst (eds)
© 2008 Taylor & Francis Group, London, ISBN 978-0-415-47535-8

Monitoring the 352 meter long Monaco floating pier

M. de Wit & G. Hovhanessian
Advitam, Vélizy, France

ABSTRACT: Again limits have been pushed further with the realization of the key element of the extension of Condamine port at Monaco, a 352 m long and 163 000 tons semi-floating pier.

The highly pre-stressed reinforcement concrete structure with a design life of 100 years is attached to the main land abutment with a very complex and 770 tons steel ball-joint system while the other end of the pier it is secured with two sets of fixed anchor chains to the seabed.

This exceptional project is a mix of building techniques, mechanical engineering, and offshore works: it includes several world records and, particularly, the spectacular connection of the ball joint system.

All these design breaking records are possible thanks to the evolutions in civil design & construction methods. In this context another evolution is of great help to allow confirming that the structures are behaving like expected by the calculation models: the monitoring tools. New technologies for the monitoring of structures are powerful tools to better understand the behavior and make sure that structure remains in good health over time.

In this paper we will review the structural health monitoring system that is installed for this extraordinary structure.

Tailor Made Concrete Structures – Walraven & Stoelhorst (eds)
© 2008 Taylor & Francis Group, London, ISBN 978-0-415-47535-8

Cambers of prestressed precast bridge girders, prediction vs. reality

M. Chandoga & A. Jaroševič
Projstar – Pk, Ltd, Bratislava, Slovakia

J. Halvonik & A. Pritula
Slovak University of Technology, Bratislava, Slovakia

P. Pšenek
Doprastav, a.s., Bratislava, Slovakia

ABSTRACT: This paper deals with investigation of reasons of deviations of actual cambers from predicted values that were observed on precast bridge girders where prestress transfer is divided into two stages. The first part of prestressing is introduced by pre-tensioned strands and usually starts only 18 hours after casting and the second part by three post-tensioned curved tendons stressed one or two month later. The problem was analyzed from structural and technological aspects.

Tailor Made Concrete Structures – Walraven & Stoelhorst (eds)
© 2008 Taylor & Francis Group, London, ISBN 978-0-415-47535-8

Monitoring of electrically isolated post-tensioning tendons

B. Elsener
ETH Zurich, Institute for Building Materials, Zurich, Switzerland

ABSTRACT: Electrically Isolated Tendons (EIT) have been introduced as one possible solution to reach the highest protection level (PL3) in the framework of *fib* recommendation for grouted post-tensioned tendons. This approach allows to check the integrity of the plastic duct during and after construction and to monitor the corrosion protection of the high-strength steel during the whole service life with electrical impedance measurements. The paper presents results on PC structures with EIT regarding quality control, long term monitoring and location of defects. Practical experience in Switzerland over the last six years was included in the revision of the Swiss Guideline "Measures to ensure the durability of post-tensioning tendons in bridges".

Tailor Made Concrete Structures – Walraven & Stoelhorst (eds)
© 2008 Taylor & Francis Group, London, ISBN 978-0-415-47535-8

Remote monitoring: Cost-effective and self-sufficient

T. Spuler, G. Moor & C. O'Suilleabhain
Mageba SA, Bülach, Switzerland

ABSTRACT: Monitoring systems for bridges and other structures offer many potential benefits over traditional assessment and monitoring methods. This paper describes some of the uses such monitoring systems can serve and the benefits they can offer, and the development of a typical solution, from identification of need and desired benefits to installation of the system on a structure. Some sample projects are then described to demonstrate the great range of purposes such remote monitoring systems can serve.

Tailor Made Concrete Structures – Walraven & Stoelhorst (eds)
© 2008 Taylor & Francis Group, London, ISBN 978-0-415-47535-8

Monitoring of Moscow Covered Center load-carrying structures

A.I. Zvedov & V.R. Falikman
Russian Engineering Academy

ABSTRACT: The organization of entertainment constructions monitoring of Covered Skating Center built in the City of Moscow, Russia, is considered.

Diagnosis

Tailor Made Concrete Structures – Walraven & Stoelhorst (eds)
© 2008 Taylor & Francis Group, London, ISBN 978-0-415-47535-8

Performance predictions of bridge structures based on damage pattern detection and extreme value methods

A. Strauss, R. Wendner & K. Bergmeister
University of Natural Resources and Applied Life Sciences, Vienna, Austria

U. Santa
A22 Autobrennero AG, Trento, Italy

D.M. Frangopol
Lehigh University, Bethlehem, PA, USA

ABSTRACT: Recently, novel numeric and monitoring techniques, such as fibre optic sensor systems and inverse damage detection analysis among others, are becoming more and more attractive for the long term reliability and performance assessment of engineering structures. These techniques combined with advanced software systems (SARA) and monitoring techniques, allow the continuously or discrete determination of the global reliability level of structures or structural components in time. Nevertheless, beyond these developments there are intentions to detect the structural performance by scanning the time dependent surface behavior, such as the pavement of bridges, and the associated stressors, by monitoring vehicles. In generally, these vehicles are equipped with laser sensor systems and high frequency cameras recording several properties of the surface. However, there is the requirement to supplement (a) the image recognition systems with the sensor readings, and (b) the performance level of a structure by means of a monitored surface degradation process. Therefore, the objectives of this paper are the presentation: (a) of methods for the incorporation of sensor readings in image recognition systems, (b) of inverse methods for the incorporation of surface damage processes in the performance level of structures, and (c) of extreme value based methods for the prediction of surface damage processes. The approaches will be applied to the existing Gossensass Bridge in Südtirol (IT).

Tailor Made Concrete Structures – Walraven & Stoelhorst (eds)
© 2008 Taylor & Francis Group, London, ISBN 978-0-415-47535-8

Diagnosis of the state of concrete structures after fire

E. Annerel & L. Taerwe
Laboratory Magnel for Concrete Research, Ghent University, Ghent, Belgium

ABSTRACT: Generally, concrete structures have a high fire resistance. After fire, it is of economical interest to reuse the structure after appropriate repair based on a reliable assessment of the residual strength. This paper deals with some fundamental aspects of a scientific and systematic methodology to assess the damage and to estimate the residual concrete strength on the basis of the change in colour and the crack development. This method seems to be promising, but the number of cracks and the change in colour are influenced by the test set up. Furthermore, these relationships change when the concrete ages after heating. Other methods such as water immersion, the Rebound Index and microscopy also provide an adequate basis for the assessment of the temperature in heated concrete.

Tailor Made Concrete Structures – Walraven & Stoelhorst (eds)
© 2008 Taylor & Francis Group, London, ISBN 978-0-415-47535-8

Chlorides ingress as an environmental load on Krk bridge

I. Stipanović Oslaković
Civil Engineering Institute of Croatia, Zagreb, Croatia

D. Bjegović
Civil Engineering Institute of Croatia, Zagreb, Croatia
Faculty of Civil Engineering, University of Zagreb, Zagreb, Croatia

D. Mikulić
Faculty of Civil Engineering, University of Zagreb, Zagreb, Croatia

ABSTRACT: Among many different environment actions causing deterioration of concrete, chlorides are representing one of the most dangerous attacks on the reinforced concrete structure. Chloride analysis were performed on the Krk bridge (Adriatic coast), which has been exposed to marine environment for 25 years. Surface chloride concentration is analysed in relation to height above the sea level and to the side of structural element. From measured chloride profiles apparent chloride diffusion coefficients were calculated and statistically analysed. Chloride ingress as an environmental load and structural resistance based on materials parameters and on concrete cover depth were statistically analysed and compared. After determination of load and resistance distributions, reliability function and reliability index were determined for the Krk bridge.

Tailor Made Concrete Structures – Walraven & Stoelhorst (eds)
© 2008 Taylor & Francis Group, London, ISBN 978-0-415-47535-8

TBM's backfill mortars – Overview – Introduction to Rheological Index

L. Linger, M. Cayrol & L. Boutillon
VINCI Construction Grands Projets, Rueil-Malmaison cedex, France

ABSTRACT: The aim of this article is to give an overview of the tail void grout mortars, which are used to backfill the gap between the tunnel lining and the ground, at the rear of the Tunnelling Boring Machine's (TBM) tail skin. These apparently "rustic" products are actually a key issue for a TBM's Project progress. They must fulfil several criteria more or less antagonistic, regarding both fresh and hardened mortar properties. The proposed approach is built up from the outcome of a large "know-how" gathered on many different projects in which VINCI has been involved. In particular, it will introduce an original and promising approach: the "**Rheological Index**". This index is a kind of generalized volumetric Water/Cement ratio, which easily provides fruitful and faithful indications regarding fresh mortar behaviour, but also concrete mixes. In other words, this original approach aims to put in "equation" the workability of the backfill mortars (and more generally cementitious materials). The first part of this article gives a general overview of the TBM's backfill mortars problematic. The second part details the original "Rheological Index" approach. The main characteristics of so-called semi-inert mortars, and associated laboratory trials, are then developed and illustrated by several examples of successful applications.

Tailor Made Concrete Structures – Walraven & Stoelhorst (eds)
© 2008 Taylor & Francis Group, London, ISBN 978-0-415-47535-8

Structural behavior with reinforcement corrosion

V.I. Carbone, G. Mancini & F. Tondolo
Department of Structural and Geotechnical Engineering, Turin, Italy

ABSTRACT: Structural degradation phenomena like reinforcement corrosion in concrete structures imply a consequent reduction in time of the safety level. Corrosion causes a reduction of the sectional area, ductility and strength of rebars, of the compressive strength of concrete caused by cracking and consequent concrete spalling, of the bond strength between steel and concrete. The subsequent redistribution of internal stresses induces a reduction in ductility at ultimate limit state and a variation of the deformational behavior in serviceability conditions. In a deteriorated member subjected to bending, concrete in compression reaches suddenly its limit deformation hindering to develop the full rotation capacity. In the present paper, the effect of a uniform corrosion on reinforcement is analyzed. A numerical model, able to take into account corrosion effects and to describe the structural behavior of concrete structures, is developed for this purpose.

Tailor Made Concrete Structures – Walraven & Stoelhorst (eds)
© 2008 Taylor & Francis Group, London, ISBN 978-0-415-47535-8

Filling inspection technology of grout in Japan

Takeshi Oshiro
West Nippon Expressway Company Limited, Japan

Keiichi Aoki
Central Nippon Expressway Company Limited, Japan

Mikio Hara
Nippon P.S. Company Limited, Japan

Akio Shoji
Oriental Shiraishi Company Limited, Japan

Tailor Made Concrete Structures – Walraven & Stoelhorst (eds)
© 2008 Taylor & Francis Group, London, ISBN 978-0-415-47535-8

Load capacity assessment of "Antonio Dovali Jaime" bridge using static and dynamic tests

O. Ortiz, J. Téllez & F.J. Burgos
Caminos y Puentes Federales de Ingresos y Servicios Conexos, México

A. Patrón, E. Reyes & V. Robles
Consultora Mexicana de Ingeniería S.A. de C.V. – Procesamiento de Ingeniería Estructural S.C., México

C. Cremona
Laboratoire Central des Ponts et Chaussées, France

M.E. Ruiz-Sandoval
Universidad Autonoma Metropolitana – Azcapotzalco, México

ABSTRACT: The "Antonio Dovali Jaime" Bridge is an important cable stayed bridge, located in the southeast of Mexico. The bridge has a total length of 1170 m and the main span is 288 m. It was open to traffic in 1984, and was the first cable-stayed bridge in Mexico. In order to asses the current state of the bridge an extensive program of non destructive tests was carried out; the field tests included measurements of stresses, a complete geometric survey of deformed configurations due to vehicle loads, and different series of ambient vibration measurements. The main results of the tests were: load-deformation relationships and dynamic properties (modal shapes, damping and frequencies) of first vibration modes. The results were employed to validate and improve a 3D finite element model of the bridge. The results obtained from FEM calculations were used to asses the actual state of the bridge.

Tailor Made Concrete Structures – Walraven & Stoelhorst (eds)
© 2008 Taylor & Francis Group, London, ISBN 978-0-415-47535-8

Safety appraisal of an existing bridge via detailed modelling

M. Pimentel, J. Santos & J.A. Figueiras
University of Porto, Faculty of Engineering, Portugal

ABSTRACT: In this work a prestressed concrete box girder bridge exhibiting cracking related pathologies is presented. Built in the late 70's, this was the first bridge to be built by the balanced cantilever method in Portugal. A detailed non-destructive inspection revealed several prestressing ducts exhibiting corrosion signs, sustaining the assumption of corroded prestressing steel in the bottom slab. The approach taken by the authors to explain the cracking origin and to evaluate the current safety level of the bridge is described. A nonlinear analysis model was developed to evaluate the ultimate load of the bridge taking into account the redundancy of the structural system. Large safety margins could be used and the safety to the ultimate states was accomplished even in a scenario with 50% of section loss in the prestressing cables located in the bottom slab. A maintenance intervention was nonetheless recommended in the short/medium term due to durability related issues.

Tailor Made Concrete Structures – Walraven & Stoelhorst (eds)
© 2008 Taylor & Francis Group, London, ISBN 978-0-415-47535-8

Cause and repair of detrimentally cracked beam in reinforced concrete bridge pier

M. Yoshizawa & K. Sasaki
Technology Center of Metropolitan Expressway, Tokyo, Japan

T. Usui & J. Sakurai
Metropolitan Expressway Company Limited, Tokyo, Japan

S. Ikeda
Yokohama National University, Tokyo, Japan

ABSTRACT: Routine inspection detected a crack up to 10 mm in width in reinforced concrete (RC) pier C1-1034 of the Circular 1 Route (C1) of the Tokyo Metropolitan Expressway. As an emergency repair, the overhanging beam was removed and rebuilt. A subsequent inspection into the cause of the crack found that the anchor bolts used to install the brackets after the construction of the bridge had cut the reinforcing bars of the beam. Thus, for those piers to which a bracket was installed during the same installation project, further inspections of the reinforcing bars were conducted to look for defects. As a result, defects were confirmed in the reinforcing bar of many of the inspected piers. Reinforcement using outside cables was added to one pier (C1-1039), as 80% of the cross-section of its reinforcing bar was defective.

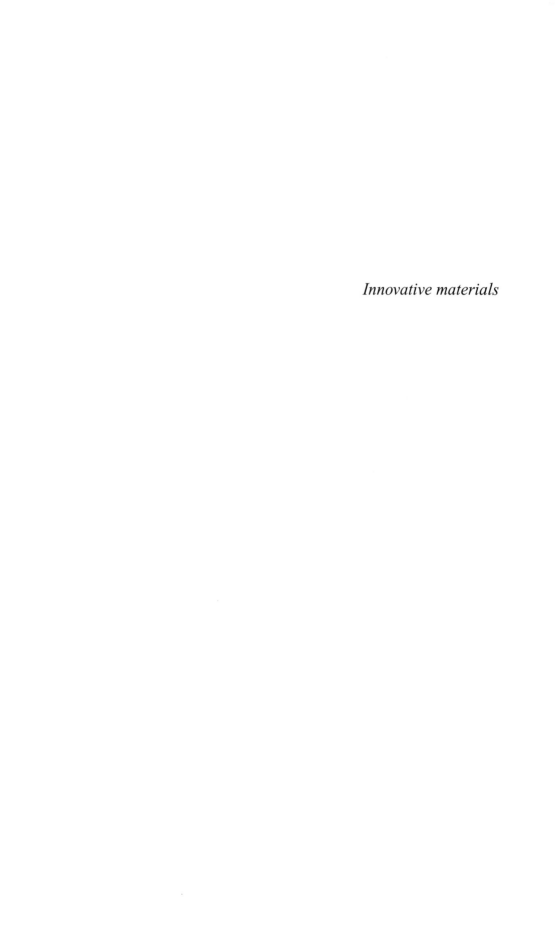

Innovative materials

Design of two reactive powder concrete bridges

M. Rebentrost
VSL Australia, Sydney, Australia

R. Annan
VSL Schweiz, Subingen, Switzerland

ABSTRACT: Ductal® is a reactive powder concrete that exhibits exceptional mechanical and durability properties. VSL has been involved with the development of this technology into structural solutions over the last ten years. Recently, concepts of two different bridges exploring the capabilities of Ductal® to the fullest have been developed. One of these is the BrennerPass footbridge located in Europe, a cable stayed structure that supports a light rail track. The other series of bridges is located in New Zealand and provides access to a variety of train stations. The design and fabrication of these two bridges is described in this paper.

Tailor Made Concrete Structures – Walraven & Stoelhorst (eds)
© 2008 Taylor & Francis Group, London, ISBN 978-0-415-47535-8

Fire resisting concrete

B.P. Van den Bossche
Concrete technology engineer

ABSTRACT: Tunnel safety is gaining importance the last few years. Fire protection is one of the items that has to protect the users of the tunnel. Effects like 'explosive scaling' occur on good quality concrete. Degeneration of the mechanical properties of concrete and rebar will also provoke dangerous instability of concrete constructions.

During the tender on the 'Overkapping A2' in Utrecht the idea of 'fire resisting concrete' was launched by the joint venture Besix-Dura Vermeer-GTI.

After acceptance of tender, a nine-month investigation was performed, starting with a literature study, to trial mixes, testing of fresh concrete and tests on hardened concrete.

Large post-tensioned specimens were made and tested on conformity with the RWS-fire curve.

After investigation the concrete was applied on the building site.

This paper will treat the whole investigation and testing of the 'fire resisting concrete', the dosing of the primary materials and the treatment of the concrete on the building site.

Tailor Made Concrete Structures – Walraven & Stoelhorst (eds)
© 2008 Taylor & Francis Group, London, ISBN 978-0-415-47535-8

Innovative ultra-high performance concrete structures

G.A. Parsekian
Federal University of São Carlos, São Carlos, Brazil

N.G. Shrive & T.G. Brown
University of Calgary, Calgary, Canada

J. Kroman
City of Calgary, Calgary, Canada

P.J. Seibert & V.H. Perry
Lafarge North-America Inc., Calgary, Canada

A. Boucher
Formerly of Cohos Evamy Integratedesign™, Calgary, Canada, now at the University of Calgary, Canada

ABSTRACT: The properties of Ductal® ultra-high performance fibre reinforced concrete are summarized together with the associated design recommendations as a basis for the use of the material in two structures in the city of Calgary – Canada. The first application was in a 20-mm thick shell canopy with no conventional reinforcement and the second was in a post-tensioned, 33.6-metre T-girder that is the central, suspended span of a pedestrian bridge. In the second application, only a small amount of glassfibre and stainless steel reinforcement was used. Both structures are believed to be milestones in the use of this type of concrete, being the first shell structure in the first instance and in the second, the largest component cast in a single pour at the time of its production. Full-scale tests on both structures were performed prior to final construction. A shell canopy was subjected to independent, full factored wind and snow load cases. The actual T-girder was subjected to 90% of the factored service load, first uniformly applied over the top of the slab and then eccentrically, over a lateral half of the slab. A summary of the design assumptions and choices, tests and results, and building processes is presented.

Tailor Made Concrete Structures – Walraven & Stoelhorst (eds)
© 2008 Taylor & Francis Group, London, ISBN 978-0-415-47535-8

Strengthening of Huisne bridge using Ultra-High-Performance Fibre-Reinforced Concrete

Thierry Thibaux

EIFFAGE TP, Neuilly sur Marne, France

ABSTRACT: Ultra-High-Performance Fibre-Reinforced Concrete (UHPFRC) is now increasingly used for innovative structures. This article describes a new application of this material to strengthen a prestressed concrete bridge near Le Mans (France).

Thin strips of UHPFRC were cast in situ along the webs of the beams, allowing the works to be performed more cheaply and faster than by conventional methods.

Tailor Made Concrete Structures – Walraven & Stoelhorst (eds)
© 2008 Taylor & Francis Group, London, ISBN 978-0-415-47535-8

Ductal® Pont du Diable footbridge, France

Mouloud Behloul
Lafarge Ciments, Saint Cloud, France

Rudy Ricciotti
Agence Rudy Ricciotti Architecte

Romain Fabio Ricciotti
Romain Fabio Ricciotti Ingénierie

Pierre Pallot
Bonna Sabla

Jacky Leboeuf
Freyssinet

ABSTRACT: Ductal®, is a new material technology developed over the last decade that offers a combination of superior technical characteristics including strength, ductility and durability, while providing highly moldable products with a quality surface. Compressive strengths reach up to 200 MPa and flexural strengths reach up to 40 MPa.

Ductal® covers a range of formulations that can be adapted to meet specific demands of different customer segments, enhancing the usage value and contributing to the overall construction performance. By utilizing the material's unique combination of superior properties, designs can eliminate passive reinforcing steel and experience reduced global construction costs, formworks, labour and maintenance; relating to benefits such as improved site construction safety, speed of construction, extended usage life and others. A number of references on Ductal® already exist in different countries, both in structural and architectural segments. In this paper, the first use of Ductal® for footbridges in France is presented.

In Gorges de l'Herault, next to Montpellier, the 'Pont du Diable' Ductal® footbridge, designed by the architect Rudy Ricciotti was constructed. This footbridge with a span of 68 m was constructed by Freyssinet and Bonna Sabla. The transversal section is composed by two bone-shaped webs, with a height of 1,8 m, connected by a light ribbed deck. Fifteen 4,6 m length segments were prefabricated in a factory are delivered to jobsite. These elements were assembled by prestressing using 8 cables and then erected in one day.

Tailor Made Concrete Structures – Walraven & Stoelhorst (eds)
© 2008 Taylor & Francis Group, London, ISBN 978-0-415-47535-8

Shear carrying capacity of Ultra-High Performance Concrete beams

Josef Hegger & Guido Bertram
Institute of Structural Concrete at RWTH Aachen University, Aachen, Germany

ABSTRACT: Ultra-High Performance Concrete (UHPC) is a high-tech material opening new opportunities especially for slender constructions. Within the collaborative research project "Sustainable Building with Ultra High Performance Concrete (UHPC)" supported by the German Research Foundation (DFG) design models for pretensioned beams have been developed at the Institute of Structural Concrete at RWTH Aachen University. Several tests were performed to investigate the bond anchorage and the shear carrying behavior of prestressed concrete beams made of UHPC with or without web openings.

Tailor Made Concrete Structures – Walraven & Stoelhorst (eds)
© 2008 Taylor & Francis Group, London, ISBN 978-0-415-47535-8

High performance materials – Advances in composite constructions

Josef Hegger & Sabine Rauscher
Institute of Structural Concrete, RWTH Aachen University, Aachen, Germany

ABSTRACT: This contribution summarizes the results of a research program (DFG priority program SPP 1182) involving the testing of push-out specimens and composite beams with innovative continuous shear connectors in ultra-high performance concrete. In the push-out tests the load-slip behavior of the shear connectors was evaluated and compared for various parameters. The parameters included the steel fiber content and orientation, as well as the transverse reinforcement ratio. The test results indicate that the steel fiber reinforcement ratio has only a minor influence on the load carrying capacity whereas the transverse reinforcement ratio is more vital. However, the fiber orientation, which is influenced by the casting direction, affects the ultimate load of the shear connector. The beam tests are focused on the moment carrying capacity and the transition of the shear forces across the joint between steel and concrete. The results of two tests are presented and discussed.

Tailor Made Concrete Structures – Walraven & Stoelhorst (eds)
© 2008 Taylor & Francis Group, London, ISBN 978-0-415-47535-8

Textile Reinforced Concrete – Realization in applications

J. Hegger, M. Zell & M. Horstmann
Institute of Structural Concrete, RWTH Aachen University, Aachen, Germany

ABSTRACT: Textile Reinforced Concrete (TRC) is a composite material made of open-meshed textile structures and a fine-grained concrete. Comparable to steel reinforcement the textile fabric bears the tensile forces released by the cracking of the concrete. Only a minimal concrete cover is required for the bond of the textile fabrics. Thus, the application of TRC leads to the design of filigree and lightweight concrete structures with high durability and high quality surfaces. In recent years, TRC has been successfully employed for the production of ventilated façade systems. Current investigations enlarge the application range of TRC to façade systems with large spans and load-bearing structures. In this paper, the investigations on self-supporting and structural sandwich panels regarding production methods, results of bending and shear tests, tests on sound insulation and fire resistance as well as first prototypes of slender frames and shell elements are presented.

Subsequent sealing of buildings made of textile reinforced concrete

R. Mott & W. Brameshuber
Institute of Building Materials Research (ibac) of RWTH Aachen University, Aachen, Germany

ABSTRACT: Many regions in Germany show a rising groundwater level. Hence the load case of buildings concerned changes from non-pressing to pressing water. Residential buildings not designed for the load case of pressing water have to be refitted. Conventional sealing methods are often associated with high complexity and high costs as well as the loss of living space. Furthermore in many cases they do not consider the additional static load of pressing water at all. This paper presents a newly developed subsequently applied sealing against pressing water. It is made of textile reinforced concrete. Using this composite material it is possible to produce a sealing system with a wall thickness of about 30 to 35 mm. During the production of an exhibit wall it became apparent that the spraying technique is an adequate and practicable method to produce a subsequent sealing of textile reinforced concrete.

Tailor Made Concrete Structures – Walraven & Stoelhorst (eds)
© *2008 Taylor & Francis Group, London, ISBN 978-0-415-47535-8*

Optimized High-Performance Concrete in Grouted Connections

S. Anders
Bilfinger Berger AG, Mannheim, Germany

L. Lohaus
Institute of Building Materials, University of Hannover, Hannover, Germany

ABSTRACT: Grouted Connections are known in the offshore industry for a long time. In the last about ten years their technology has been projected for the foundations of Offshore Wind Energy Converters with the difference that High-Performance Concrete is applied. One of the striking advantages of Grouted Connections is the equal load-bearing capacity in tension and compression. It is shown that an increasing compressive strength of the grout, increasing shear key height and fiber reinforcement increase the load-bearing capacity of Grouted Connections. At the same time the danger of shear failure of the shear keys and yielding in the steel members raises. Equations for an appropriate design of the steel members are given, depending on the compressive strength of the grout. Thus, High-Performance Concrete for Grouted Connections can be specifically optimized for the demands of the joint. Finally, further applications e.g. for joining steel members or tower-like structures basing on Grouted Connections are presented, which suggest Grouted Connections as a joining technology for the future.

Tailor Made Concrete Structures – Walraven & Stoelhorst (eds)
© 2008 Taylor & Francis Group, London, ISBN 978-0-415-47535-8

Tailored superplasticisers for tailor made concrete structures

Mario Corradi, Rabinder Khurana, Roberta Magarotto & Sandro Moro
BASF Construction Chemicals, Admixtures Systems Europe, Treviso, Italy

ABSTRACT: Concrete is a versatile material and offers new opportunities for developing innovative structural forms. The last few years have shown revolutionary developments of new types of concrete, specially self-compacting concrete, whose use is growing rapidly in Europe, and fibre reinforced ultra high performance concrete whose applications are still in the very preliminary stages. The extraordinary properties of Ultra High Performance Steel Fibre Reinforced Concrete makes it an excellent alternative to steel as a construction material and allows the construction of sustainable and cost effective structures. Recommendations and guidelines on these materials are available but International Standards will have to be developed.

In this paper, results of a study made to develop "user friendly" UHPC mixtures are presented. These mixes have constituents that are readily available locally and the mixing and the casting procedures as similar to the existing ones. Thanks to the development of innovative superplasticisers, whose mechanism of action is illustrated, the mixtures are self-compacting with high flow and adequate plastic viscosity. The workability time can be tailored for either precast applications or cast "in situ" structures.

Tailor Made Concrete Structures – Walraven & Stoelhorst (eds)
© 2008 Taylor & Francis Group, London, ISBN 978-0-415-47535-8

Jointless prestressed concrete viaduct using ECC

M. Fujishiro & K. Suda
Kajima Corporation, Tokyo, Japan

Y. Nagata
Metropolitan Expressway Corporation, Tokyo, Japan

ABSTRACT: On concrete superstructure consisting with continuous multiple simple girders, the surface level difference at expansion joint has been disturbing drivers' comfort. And it has been causing vibration and making noise in surrounding areas. To solve the problems, the authors developed a jointless method that provides for quick construction by connecting separate girders on both sides of the joint clearance using plates made of ECC (Engineered Cementitious Composite) to form continuous pavement. This paper describes the material properties of ECC, an outline of the developed method, results of a verification tests, report of field practice and development in the future.

Tailor Made Concrete Structures – Walraven & Stoelhorst (eds)
© *2008 Taylor & Francis Group, London, ISBN 978-0-415-47535-8*

Pumping of Self Compacting Concrete: An insight into a daily application

D. Feys
Magnel Laboratory for Concrete Research, Department of Structural Engineering, Faculty
of Engineering, Ghent University, Ghent, Belgium
Hydraulics Laboratory, Department of Civil Engineering, Faculty of Engineering, Ghent University, Ghent, Belgium

R. Verhoeven
Hydraulics Laboratory, Department of Civil Engineering, Faculty of Engineering, Ghent University, Ghent, Belgium

G. De Schutter
Magnel Laboratory for Concrete Research, Department of Structural Engineering,
Faculty of Engineering, Ghent University, Ghent, Belgium

ABSTRACT: Self-Compacting Concrete (SCC) is nowadays a worldwide applied material in construction. Due to its self-compactability, a large reduction in compaction energy and noise can be established, also making the resulting structure less dependent on the skills of the workmen on site. SCC is placed in the formwork in similar ways as for traditional concrete, meaning that pumping of SCC occurs in practical situations. Due to the absence of the need for compaction, SCC can even be pumped from the bottom of the formwork.

Taking into account the results of rheometer tests on different kinds of SCC, high pressures are needed to make SCC flow in pipes at a reasonable casting rate, if no slippage or segregation occurs. In this paper, a comparison between rheometer tests, flow tests at very low discharges and full scale pumping tests is described.

Tailor Made Concrete Structures – Walraven & Stoelhorst (eds)
© 2008 Taylor & Francis Group, London, ISBN 978-0-415-47535-8

Replacement of shear reinforcement by steel fibres in pretensioned concrete beams

P. De Pauw & L. Taerwe
Magnel Laboratory for Concrete Research, Department of Structural Engineering, Faculty of Engineering, Ghent University, Ghent, Belgium

N. Van den Buverie & W. Moerman
Willy Naessens Industriebouw nv., Wortegem-Petegem, Belgium

ABSTRACT: By means of loading tests up to failure, the shear behaviour of precast pretensioned concrete beams made with steel fibre concrete and without conventional shear reinforcement is compared with the shear behaviour of a standard beam made with concrete without fibres but with stirrups as shear reinforcement. A beam made of plain concrete without shear reinforcement is used to investigate the effect of both types of shear reinforcement. The beams are designed according to Eurocode 2 and, especially for the steel fibre concrete beams, according to the guidelines of the "σ − ε − design method" recommended by the RILEM TC 162-TDF technical committee and according to information found in literature. From the test results it can be concluded that, for the beams considered, ordinary shear reinforcement can be eliminated by using steel fibre reinforced concrete. However, due attention should be paid to the mixing and casting procedures of the steel fibre concrete.

Tailor Made Concrete Structures – Walraven & Stoelhorst (eds)
© 2008 Taylor & Francis Group, London, ISBN 978-0-415-47535-8

Properties and applications of DUCON® A micro-reinforced ultra-high-performance concrete

Jens Schneider & Jörg Reymendt

Frankfurt University of Applied Sciences, Department of Architecture and Civil Engineering,
Frankfurt, Germany

ABSTRACT: The main advantage of DUCON (DUctile CONcrete), a patented 3-D micro-reinforced, self-compacting Ultra High Performance Concrete, is its ductility in combination with high compressive and flexural strength. This paper describes the properties and recent applications of DUCON. The material proved to have exceptional advantages as explosion resistant and bullet/fragment proof material. Explosion tests performed showed high energy absorption and ductility, i.e. no failure and no fragment projectiles. Special sniper projectiles which easily penetrate standard reinforced concrete can be stopped by DUCON panels. The main field of applications and projects realized so far are therefore the protection of critical infrastructure, such as military and public buildings, embassies, power plants, banks and data centers and ammunition storage tanks. Here, DUCON is used in new structures or for retro-fitting of existing structures in slabs and walls (fragmentation protection), for column protection, protective elements, etc. Moreover, other engineering and architectural projects realized such as industrial overlay, retro-fitting of columns for earthquake protection, slim façade panels and interior design elements show the great variety in applications of the material.

Tailor Made Concrete Structures – Walraven & Stoelhorst (eds)
© 2008 Taylor & Francis Group, London, ISBN 978-0-415-47535-8

Deformation behavior of reinforced UHPFRC elements in tension

V. Sigrist & M. Rauch

Institute of Concrete Structures, Hamburg University of Technology (TUHH), Hamburg, Germany

ABSTRACT: Ultra High Performance Fiber Reinforced Concrete (UHPFRC) is characterized by a compressive strength in the range of 200 MPa as well as by improved durability properties compared to normal strength concrete. The addition of short steel fibers influences the behavior of UHPFRC positively, but to attain ductile failures conventional reinforcement still is required. The mechanical properties of UHPFRC have been widely studied. However, in practical application UHPFRC is quite unexplored. Therefore, the deformation behavior of UHPFRC tension ties reinforced with conventional as well as high-strength steel has been examined. In this paper the results of an experimental study are presented. Based on these, an analytical model is developed which enables the calculation of the cracking and deformation behavior of such elements. The overall aim of this research is to develop reliable methods for the conception and design of structural members made of UHPFRC.

Tailor Made Concrete Structures – Walraven & Stoelhorst (eds)
© 2008 Taylor & Francis Group, London, ISBN 978-0-415-47535-8

Structural performance of pretensioned member with Ultra High Strength Fiber Reinforced Concrete

T. Ichinomiya, N. Sogabe, Y. Taira & Y. Hishiki
Kajima Technical Research Institute, Kajima Corporation, Tokyo, Japan

ABSTRACT: The authors have developed a kind of Ultra High Strength Fiber Reinforced Concrete (UFC) with high compressive strength and high tensile ductility and have been studying for practical use. UFC makes it possible to reduce the weight of prestressed concrete structures. Particularly in case that UFC is applied to pretensioned members, high bond strength can also be expected to reduce the transfer length. Autogenous shrinkage and creep in early age of UFC, however, reduces effective prestress. In this study, its basic mechanical properties concerning effective prestress such as shrinkage and creep in early age were investigated to get design values for the material. Prestressing tests were also conducted to determine transfer length and effective stress. Furthermore, flexural and shear tests using beam members were conducted and it was revealed that flexural and shear capacity could be estimated using the formula shown in "Recommendations for Design and Construction of Ultra High Strength Fiber Reinforced Concrete Structures, -Draft" by Japan Society of Civil Engineers.

Tailor Made Concrete Structures – Walraven & Stoelhorst (eds)
© 2008 Taylor & Francis Group, London, ISBN 978-0-415-47535-8

Measuring the packing density to lower the cement content in concrete

S.A.A.M. Fennis & J.C. Walraven
Delft University of Technology, Delft, The Netherlands

T.G. Nijland
TNO Built Environment & Geosciences, Delft, The Netherlands

ABSTRACT: Geometrically based particle packing models can help to predict the water demand of concrete, and thus the material properties. In this paper it is described how centrifugal consolidation can be used to determine the packing density of powders. The method is assessed based on experimental data, calculations and polarization and fluorescence microscopy of the samples. Results show that an average maximum packing density can be measured, which depends on the initial water powder ratio, the use of superplasticizer, the mixing procedure of the paste and the applied compaction energy. Viscosity measurements show the influence of the particle packing density on water demand and how concrete mixtures can be designed to lower the cement content in concrete.

Development of a bacteria-based self healing concrete

Henk M. Jonkers & Erik Schlangen
Delft University of Technology, Faculty of Civil Engineering and GeoSciences/Microlab,
Delft, The Netherlands

ABSTRACT: Concrete structures usually show some self-healing capacity, i.e. the ability to heal or seal freshly formed micro-cracks. This property is mainly due to the presence of non-hydrated excess cement particles in the materials matrix, which undergo delayed or secondary hydration upon reaction with ingress water. In this research project we develop a new type of self-healing concrete in which bacteria mediate the production of minerals which rapidly seal freshly formed cracks, a process that concomitantly decreases concrete permeability, and thus better protects embedded steel reinforcement from corrosion. Initial results show that the addition of specific organic mineral precursor compounds plus spore-forming alkaliphilic bacteria as self-healing agents produces up to 100-μm sized calcite particles which can potentially seal micro- to even larger-sized cracks. Further development of this bio-concrete with significantly increased self-healing capacities could represent a new type of durable and sustainable concrete with a wide range of potential applications.

Tailor Made Concrete Structures – Walraven & Stoelhorst (eds)
© *2008 Taylor & Francis Group, London, ISBN 978-0-415-47535-8*

Use of polypropylene fibres to reduce explosive spalling in concretes exposed to fire

J.A. Larbi, A.J.S. Siemes & R.B. Polder
TNO Built Environment & Geosciences, Delft, The Netherlands

ABSTRACT: Explosive spalling of pieces of concrete from the heated surface is considered to the most dangerous effect of damage of concrete subjected to intense fire attack, especially when it occurs in restricted areas such as underground tunnels. Recent investigations have revealed that the amount of explosive spalling and the extent of cracking can considerably be reduced by use of suitable amount of polypropylene fibres. However, little attention has been given to exactly how the fibres behave in concrete matrix when exposed to fire. A good insight into the behaviour of the fibres when applied in concrete, especially when subjected to fire, can help optimise their use to reduce explosive spalling. This paper deals with a case study, in which an integrated microscopic method was used to assess the effectiveness of the pp-fibres in reducing explosive spalling in concrete elements subjected to fire attack. RCM test was also performed on standard specimens to establish whether the presence of the fibres adversely affect the permeability and durability of the elements.

Tailor Made Concrete Structures – Walraven & Stoelhorst (eds)
© *2008 Taylor & Francis Group, London, ISBN 978-0-415-47535-8*

Steel fibre only reinforced concrete in free suspended elevated slabs: Case studies, design assisted by testing route, comparison to the latest SFRC standard documents

Xavier Destrée
Consultant, Structural Engineer, La Hulpe, Belgium

Jürgen Mandl
Research and Development ArcelorMittal Bissen, Bissen, Luxembourg

ABSTRACT: The total replacement of traditional rebars is now completely routine for 15 year in applications such as industrial and commercial suspended slabs resting on pile grids, which can span from 3 m to 5 m each way, with span to depth ratios from 15 to 20. Seven millions of square meter have been completed so far. More recently, the structural use of steel fibre-only reinforcement at high dosage rate has been developed as the sole method of reinforcement for fully elevated suspended slabs spanning from 5 m to 8 m each way, with a span to depth ratio of 30. More than forty buildings are now completed. The SFR concrete mix is also fully pumpable and doesn't need any poker vibrating. Significant time and cost savings are then achieved. Design methods are derived from round slabs flexion testing and from full scale testing results of real elevated suspended slabs. Three full scale testing slabs have been built in order to monitor deflections, punching and cracking when they are loaded up to final rupture. This article summarizes the design methods and compares them to the provisions of the most recently available steel fibre reinforced concrete standards.

Tailor Made Concrete Structures – Walraven & Stoelhorst (eds)
© 2008 Taylor & Francis Group, London, ISBN 978-0-415-47535-8

Evaluation of the bond strength behavior between steel bars and High Strength Fiber Reinforced Self-Compacting Concrete at early ages

F.M. Almeida Filho, M.K. El Debs & A.L.H.C. El Debs
Department of Engineering Structures, University of São Paulo, São Carlos, Brazil

ABSTRACT: Self-Compacting Concrete (SCC) appeared to avoid the difficult and onerous work of concrete vibration. It can be defined as a material that is capable to flow inside of a formwork, passing by the reinforcement and filling out the formwork, without the use of any vibration equipments. The main objective of this research was developing a high strength SCC with steel fibers for use in precast connections and also to evaluate the behavior of its bond strength with the steel bars of different diameters. The study considered as fundamental parameters, the steel fibers and the steel bars diameter. According to the results, the incorporation of steel fibers increased the bond strength of SCC giving it a ductile behavior according to the fiber content and fiber length. The development of a HSFRSCC reveal itself ideal for use in precast connections as a filling material, producing a more monolithic behavior for the precast structures.

Tailor Made Concrete Structures – Walraven & Stoelhorst (eds)
© 2008 Taylor & Francis Group, London, ISBN 978-0-415-47535-8

Research on the cracking control and pumpability of HPC in S-C segment of Sutong Bridge

H. Zhang, S.K. Li & J.F. Tao

The Second Bureau of Harbor Engineering, CCCG, Wuhan, Hubei, China

ABSTRACT: The height of the main tower of Sutong Bridge is 306 meters and the strength grade of concrete casting cover the steel anchor box is C50. The Elastic Modulus between steel and concrete is different obviously, so the concrete is prone to cracking, thus the pumpability and cracking-resisting of the concrete cause much attention. In this paper, fly ash, hybrid fibers (steel fibers and polypropylene fibers), water content, and polycarboxylic water-reducing agent were used to optimize the mix proportion of concrete. The elementary mechanical properties of concrete including compressive strength, flexural strength, split tensile strength and elastic modulus were investigated and the cracking-resisting property of different concrete mix proportions was tested. The results indicated that the mechanical properties of concrete satisfied the construction requirement, the cracking–resisting property of concrete mixed with hybrid fibers was the best and the optimized concrete was pumped to the height o in site.

Tailor Made Concrete Structures – Walraven & Stoelhorst (eds)
© 2008 Taylor & Francis Group, London, ISBN 978-0-415-47535-8

Development strategies for foamed cement paste

J.U. Pott
Beton Marketing Nord GmbH, Hannover, Germany

L. Lohaus
Institut für Baustoffe, Leibniz Universität Hannover, Hannover, Germany

ABSTRACT: For several decades numerous research projects dealt with foamed concrete. Although foamed cement-bound materials have very useful properties, for example low density and low thermal conductivity, they are not often used as construction material, because predefined properties are difficult to attain accurately. Therefore the intention of this research work is the unerring production of cement-bound foams. Based on technical and scientific basics of foam technology and concrete technology, special attention is paid to the rheological behaviour, mixture stability and the hardening process of cement paste as the dominating factors for its optimisation as liquid phase for foamed cement paste. From these considerations a comprehensive model for stable cement-bound foams is developed. Within this model the complex interaction of different components of cement pastes and their influences during the phases of production, workmanship and hardening are described. According to the before mentioned aspects, a stable composition for a foamed cement paste was developed. Modifications of this composition have been tested, in order to prove the reliability of the theoretical model.

Tailor Made Concrete Structures – Walraven & Stoelhorst (eds)
© 2008 Taylor & Francis Group, London, ISBN 978-0-415-47535-8

Expanding the application range of RC-columns by the use of UHPC

M. Empelmann, M. Teutsch & G. Steven
Institute for Building Materials, Concrete Construction and Fire Protection (iBMB), Technical University of Braunschweig, Braunschweig, Germany

ABSTRACT: The development of Ultra-High Performance Concrete (UHPC) with compression strengths up to 200 MPa widens the application range for RC-constructions. But UHPC shows, in comparison to normal and high-strength concrete, a brittle material behaviour when the ultimate load-bearing capacity is reached. Experimental research carried out at the iBMB of the Technical University of Braunschweig, Germany, shows that the load-bearing behaviour of UHPC-columns can be improved considerably by the addition of steel fibres, leading to Ultra High Performance Fibre Reinforced Concrete (UHPFRC).

On this basis, it is now possible to use UHPFRC-columns for constructions, which so far have been reserved for steel and/or composite constructions. By the use of UHPFRC the load-bearing capacity of RC-columns can be adjusted in a "tailor-made" way, according to the individual loading situation.

This paper will present results of several numerical and experimental studies, with regard to the use of UHPFRC-columns in comparison to normal or high-strength concrete columns and composite columns.

Tailor Made Concrete Structures – Walraven & Stoelhorst (eds)
© 2008 Taylor & Francis Group, London, ISBN 978-0-415-47535-8

Quality concrete surfaces mean a longer life asset

D.J. Wilson
Max Frank GmbH, Leiblfing, Germany

ABSTRACT: To minimize operational and maintenance costs, concrete surfaces in direct contact with water must have: resistance to chemical and physical attack, microbiological growth and mechanical load and remain watertight. Several factors which are generally ignored when assessing how to achieve quality durable watertight concrete are formwork type, spacers, release agent usage and curing. Many tanks have to be rehabilitated even before commissioning as their concrete surfaces do not comply with the above requirements due to biofilm development, surface depassivation due to water borne CO_2 and the overall poor quality of the covercrete zone. Repairs are expensive, so avoidance of these problems by adopting a holistic approach to tank construction is essential. The benefits deriving from the use of high quality construction accessories such as CPF liners and fibre concrete spacers and distance tubes will improve asset quality and life.

Tailor Made Concrete Structures – Walraven & Stoelhorst (eds)
© 2008 Taylor & Francis Group, London, ISBN 978-0-415-47535-8

Textile Reinforced Concrete (TRC) for precast Stay-in-Place formwork elements

I.C. Papantoniou
Civil Engineering Department, University of Patras, Patras, Greece

C.G. Papanicolaou
Structural Materials Laboratory, Civil Engineering Department, University of Patras, Patras, Greece

ABSTRACT: The main goal of the present study is to experimentally investigate the response of structural elements cast against thin-walled stay-in-place formwork elements made of Textile Reinforced Concrete (TRC). TRC comprises an innovative composite material consisting of fabric meshes made of long woven, knitted or even unwoven fibre yarns (e.g. carbon, glass or aramid) in at least two (typically orthogonal) directions embedded in a cementitious matrix (mortar or fine-grained concrete). The experimental investigation described in this study was carried out on two types of reinforced concrete specimens: the first one included 22 beam-type specimens incorporating flat TRC stay-in-place formworks and the second one included 11 prismatic column-type specimens cast into permanent precast TRC shafts.

Tailor Made Concrete Structures – Walraven & Stoelhorst (eds)
© 2008 Taylor & Francis Group, London, ISBN 978-0-415-47535-8

Bridges utilizing high strength concrete

J. Strasky, I. Terzijski & R. Necas
Brno University of Technology, Faculty of Civil Engineering, Brno, Czech Republic

ABSTRACT: Possibilities of structures utilizing high strength concrete are shown on examples of several highway and pedestrian bridges recently built in the Czech Republic. These structures are discussed from the point of view of their architectural and structural solution. Also a process of their erection is presented.

Codes for the future

Tailor Made Concrete Structures – Walraven & Stoelhorst (eds)
© *2008 Taylor & Francis Group, London, ISBN 978-0-415-47535-8*

Modelling of shear-fracture of fibre-reinforced concrete

G.G. Lee & S.J. Foster

The University of New South Wales, Australia

ABSTRACT: For design, engineers require a simple yet reliable approach that explains and models with sufficient accuracy the behaviour under load of fibre reinforced concrete. In this paper, the Variable Engagement Model II is developed to describe the behaviour of randomly orientated steel-fibre reinforced composites subject to mode II (shear) fracture. The model is developed by integrating the behaviour of single, randomly oriented, fibres over 3D space and, based on quantifiable material and mix parameters, is shown to be capable of capturing the stress versus crack sliding displacement response for steel-fibre reinforced composites in both the pre- and post-peak stages.

Tailor Made Concrete Structures – Walraven & Stoelhorst (eds)
© 2008 Taylor & Francis Group, London, ISBN 978-0-415-47535-8

The American P2P initiative

Ken W. Day
Consultant

ABSTRACT: P2P is an acronym for Prescription to Performance and relates to whether concrete should be specified by how it is to be produced, or by what properties it should have. The purpose of this paper is to report on progress on this initiative, to summarise its origins and benefits, and to examine the remaining obstacles.

Tailor Made Concrete Structures – Walraven & Stoelhorst (eds)
© 2008 Taylor & Francis Group, London, ISBN 978-0-415-47535-8

Dispersion of the mechanical properties of FRC investigated by different bending tests

B. Parmentier & E. De Grove
Belgian Building Research Institute, Limelette, Belgium

L. Vandewalle & F. Van Rickstal
Catholic University of Leuven, Leuven, Belgium

ABSTRACT: Fibres are well known to provide post-cracking energy to concrete, enhancing the energy absorption capacity (toughness) of the material. However, variations of toughness performances of FRC are a well-known limitation for its use in structural applications. This shortcoming is not only caused by the heterogeneity of the material itself. It is also emphasized by the test method used to determine the FRC properties. This paper presents the results of an extended test programme on FRC. Four testing methods are compared. Special attention is paid to the analysis of the variation on the results since this variation indicates the power of discernment of a test method. A strong correlation ($R^2 > 0.99$) has been found between the results of the round panel test and the RILEM beam test for the same deflection/cracking level.

BA-Cortex: Learning tools for EC2

C. Lanos & V. Bonamy
IUT Université Rennes 1, Rennes, France

P. Guiraud
CIMBETON, Paris, France

C. Casandjian
INSA, Rennes, France

ABSTRACT: The project named BA-Cortex relates more particularly to the "Eurocodes 2: Design of concrete structures – Part 1-1: General rules and rules for buildings". The objective of the project is to build an educative tool accessible to everyone via a website. The didactic tool presents different training, dimensioning and project modules. It synthesizes a basic teaching for reinforced concrete and pre-stressed concrete, examples of calculation and project situations.

Tailor Made Concrete Structures – Walraven & Stoelhorst (eds)
© 2008 Taylor & Francis Group, London, ISBN 978-0-415-47535-8

Shear resistance of bridge decks without shear reinforcement

G.A. Rombach & S. Latte
Institute of Concrete Structures, Hamburg University of Technology, Hamburg, Germany

ABSTRACT: This paper presents experimental and numerical investigations of a research project, carried out to examine whether bridge deck slabs under concentrated wheel loads exhibit reserves of shear capacity, which have been neglected in the current Eurocode design provisions so far. Tests conducted on large-scale cantilever slabs under concentrated loads demonstrated significant load redistributions after the formation of bending and diagonal shear cracks. The evaluation of the test results showed, that the current design formula leads to rather conservative values of shear capacity for bridge deck slabs when combined with a shear force distribution according to linear elastic slab analysis. A beneficial effect of a tapered slab bottom on the shear capacity, as implied by most codes, could not be verified with these tests.

Tailor Made Concrete Structures – Walraven & Stoelhorst (eds)
© *2008 Taylor & Francis Group, London, ISBN 978-0-415-47535-8*

Transverse flexural and torsional strength of Prestressed Precast Hollow-Core Slabs

A. Pisanty
Faculty of Civil and Environmental Engineering, Technion, Haifa, Israel

ABSTRACT: The transverse flexural and pure torsional strength of Prestressed Precast Hollow-Core Slabs (PPHCS) were investigated experimentally. Two series of tests were conducted and their results reported, relating to: a. the lateral (perpendicular to span) flexural strength, and b. the pure torsional strength. The development of a machine for the testing of slab cuts under pure torsion is presented and explained in detail. Evidence of existing reliable tensile/shear strength was produced. A size effect was detected indicating a higher strength with the reduction of the slab thickness, under both bending and torsion.

Tailor Made Concrete Structures – Walraven & Stoelhorst (eds)
© 2008 Taylor & Francis Group, London, ISBN 978-0-415-47535-8

Influence of bond-slip on the behaviour of reinforced concrete beam to column joints

M.L. Beconcini, P. Croce & P. Formichi
Department of Structural Engineering, University of Pisa, Pisa, Italy

ABSTRACT: The bond-slip correlation proposed by CEB is the most commonly adopted to predict slip displacements at the interface between steel bars and concrete. As known, the CEB model is based upon experimental results, obtained from pull-out tests performed on bars anchored by concrete for a limited length (usually $5\,\phi$). In real beams the steel bar is usually bonded for a much longer length, and even in presence of concrete cracks, the maximum slip does not generally reach the values predicted by the CEB curve. An evidence of the above has been observed analysing experimental data, registered during a wide campaign for testing full scale r.c. beam to column joints, carried out by the Authors. Consequently, it has been calibrated a modified $\tau - s$ curve and an innovative test arrangement has been proposed, able to investigate bond-slip correlation for rebars in actual anchoring conditions.

In the paper they are illustrated some promising preliminary results, obtained with the proposed testing set up, confirming the experimental evidence about the significant modification of the bond-slip curve with respect to the CEB one.

Tailor Made Concrete Structures – Walraven & Stoelhorst (eds)
© 2008 Taylor & Francis Group, London, ISBN 978-0-415-47535-8

Improvement in the plastic rotation evaluation by means of fracture mechanics concepts

M. Corrado, A. Carpinteri, M. Paggi & G. Mancini
Politecnico di Torino, Department of Structural Engineering and Geotechnics, Torino, Italy

ABSTRACT: The well-known *Cohesive Crack Model* describes strain localization with a softening stress variation in concrete members subjected to tension. Based on the assumption that strain localization also occurs in compression, the *Overlapping Crack Model*, analogous to the cohesive one, is proposed to simulate material compenetration due to crushing. By applying this model, it is possible to describe the size effects in compression in a rational way. The two aforementioned elementary models are then merged into a more complex algorithm based on the finite element method, able to describe both cracking and crushing growths during loading processes in RC members. With this algorithm in hand, it is possible to investigate on the influence of the reinforcement percentage and/or the structural size of RC beams, with special attention to their rotational capacity. The obtained results evidence that the prescriptions concerning the plastic rotations provided by codes of practice, not taking into account the scale effects, are not conservative in the case of large structural sizes.

Tailor Made Concrete Structures – Walraven & Stoelhorst (eds)
© 2008 Taylor & Francis Group, London, ISBN 978-0-415-47535-8

Reliability-based code calibration and features affecting probabilistic performance of concrete bridge

I. Paik
Kyungwon University, Kyunggi-Do, Korea

S. Shin
Inha University, Incheon, Korea

C. Shim
Chung-Ang University, Seoul, Korea

ABSTRACT: This paper presents some of the features during reliability based code calibration process for the newly proposed Korean bridge design code. Field data for the material strengths of concrete and reinforcing steel are collected and the statistical properties are obtained. Partial safety factors for material strengths and member forces can be recommended after calibration for target reliability index. Sensitivity of design variables and distribution type to reliability index is examined.

Tailor Made Concrete Structures – Walraven & Stoelhorst (eds)
© 2008 Taylor & Francis Group, London, ISBN 978-0-415-47535-8

Codes for SFRC structures – A Swedish proposal

J. Silfwerbrand
Swedish Cement and Concrete Research Institute, Stockholm, Sweden

ABSTRACT: Despite both technical and economical advantages, the use of Steel Fibre Reinforced Concrete (SFRC) has been limited to certain areas, e.g., shotcrete for rock strengthening and slabs-on-grade. In order to facilitate its design and increase its use, design recommendations are necessary. The Swedish Concrete Association has tried to promote the use of SFRC by issuing recommendations both for general applications and for industrial concrete floors. This paper summaries the Swedish proposal to a possible Code for SFRC structures. It deals with basis of design, determination of material properties, slabs-on-grade, pile-supported slabs, and overlays.

Tailor Made Concrete Structures – Walraven & Stoelhorst (eds)
© 2008 Taylor & Francis Group, London, ISBN 978-0-415-47535-8

Shear strength in one- and two-way slabs according to the Critical Shear Crack Theory

A. Muttoni & M. Fernández Ruiz
Ecole Polytechnique Fédérale de Lausanne, Lausanne, Switzerland

ABSTRACT: Currently, there is no generally-accepted theory giving a physical explanation of the shear strength in one- and two-way slabs. Furthermore, for members without transverse reinforcement, shear strength is estimated in most codes of practice following empirical or semi-empirical approaches. In this paper, the fundamentals of the Critical Shear Crack Theory (CSCT) are introduced. This theory, based on a mechanical model, is shown to provide a unified approach for one- and two-way shear in slabs, leading to simple design expressions for estimating the strength and deformation capacity of such members. The paper also details a code-like formulation based on this theory and developed for the Swiss code for structural concrete. Comparisons of the theory to a wide range of test data are finally presented.

Tailor Made Concrete Structures – Walraven & Stoelhorst (eds)
© 2008 Taylor & Francis Group, London, ISBN 978-0-415-47535-8

Strut-and-tie modelling of short span beams

J. Sagaseta & R. Vollum

*Department of Civil and Environmental Engineering, Concrete Structures Section, Imperial College London,
London, UK*

ABSTRACT: In High Strength Concrete, (HSC) aggregate may fracture leading to relatively smooth cracks
and a possible reduction in shear strength. There seems little doubt that shear strength is reduced in beams without
stirrups due to aggregate fracture. The influence of aggregate fracture is less clear in beams with stirrups due to a
lack of appropriate test data. This paper presents results from a series of 16 beam tests which were carried out at
Imperial College London to assess the influence of aggregate fracture on the shear strength of beams with a ratio
of shear span to effective depth of either 1.5 or 3.5. This paper focuses on the short span beams. A simple strut
and tie model is presented for the analysis of short span beams that is consistent with the design assumptions
in EC2. The model is validated with results from 214 tests and is shown to give good predictions of shear
strength.

Tailor Made Concrete Structures – Walraven & Stoelhorst (eds)
© *2008 Taylor & Francis Group, London, ISBN 978-0-415-47535-8*

Modelling of shrinkage induced curvature of cracked concrete beams

R. Mu, J.P. Forth & A.W. Beeby
The University of Leeds, Leeds, UK

R. Scott
University of Durham, Durham, UK

ABSTRACT: Besides load and temperature, shrinkage and creep are the main factors that influence the curvature of reinforced concrete sections. For cracked sections, this effect is calculated semi-empirically in structural design codes such as BS8110 and EC2. In order to verify the accuracy of the calculation, the curvature of cracked section of concrete beams due to shrinkage was analyzed numerically and validated experimentally. The analysis in this paper is based on the mechanical equilibrium method and with basic assumptions of plane sections remaining plane and linear creep superposition. This method divides a section into a number of strips. The neutral axis position and curvature of the section were determined by iteration until equilibrium was obtained. To verify the calculation results, two beams, cast with different shrinkage but similar creep concrete, were tested. In theory, any difference in curvature or deflection between the two beams was therefore caused by shrinkage. The results showed that the model proposed in this investigation for a cracked section adequately predicts the time-dependent (creep and shrinkage) curvature of the experimental beams. Comparing the shrinkage curvatures determined using the codes (EC2 and 8110) with the curvatures of the measured beams and those predicted by the model proposed in this investigation, the code methods are suitably accurate for cracked beams.

Tailor Made Concrete Structures – Walraven & Stoelhorst (eds)
© 2008 Taylor & Francis Group, London, ISBN 978-0-415-47535-8

Investigation of cracks surface roughness and shear transfer strength of cracked HSC

V.P. Mitrofanov

Poltava National Technical University, Poltava, Ukraine

ABSTRACT: The definition of *roughness characteristic of rupture crack surface (RCCS)* is introduced. The results of RCCS investigation of different strength concretes on the special experimental installation are stated. The RCCS differences of normal and high strength concretes are displayed. On the RCCS basis *the criterion of existence of interlock (CEI)* of roughness on the crack sides is introduced. The model of resistance to shear in crack is worked out.

Problems on Harmonization of Ukraine and EU Normative Bases in the Area of Concrete Structures

P. Kryvosheyev, A. Bambura & Yu. Slyusarenko
The State Research Institute of Building Constructions, Kyiv, Ukraine

ABSTRACT: Comparative analysis on basic construction regulations of Ukraine and the European Union in the area of concrete structures is presented. There is pointed that the real state of civil engineering and normative base acting in Ukraine proved the inability to accept now a Code identical to Eurocode 2 as a National design norm for concrete structures. The design Codes on concrete structures being developed in Ukraine needs to be modified at that it is necessary to provide for a transition period when the both Codes will be valid.

Tailor Made Concrete Structures – Walraven & Stoelhorst (eds)
© 2008 Taylor & Francis Group, London, ISBN 978-0-415-47535-8

Developing non linear analysis methods for codes of practice

D.A. Kuchma
University of Illinois at Urbana-Champaign

ABSTRACT: Structural Engineering design practice primarily relies upon the use of linear elastic analysis methods for determining demands on structural components and then empirical provisions for ensuring that these components have sufficient capacity to support factored loads. While this approach has well served the profession for decades, it does not enable the design for all levels of performance, leads to inconsistencies in and between codes-of-practice, and does not enable designers to take full advantage of new materials, his or her intuition, and the inelastic behavior of materials. A new non-linear design and analysis environment is needed where the designer has access to computational tools that provide quantifiably-accurate predictions of all critical aspects of performance. The creation of this environment will require the use of new experimental methods, instrumentation devices, visualization tools, data-archives of benchmark tests, model validation procedures, and support for the use of non-linear computational tools in professional practice.

Tailor Made Concrete Structures – Walraven & Stoelhorst (eds)
© 2008 Taylor & Francis Group, London, ISBN 978-0-415-47535-8

Safety with prestrained elastic restraints

D.L. Allaix, V.I. Carbone & G. Mancini
Department of Structural and Geotechnical Engineering, Politecnico di Torino, Torino, Italy

ABSTRACT: an increasing number of structures containing elastic restraints subject to imposed deformations is being built, both within the field of infrastructures and in the field of buildings. The presence of imposed deformations within the elastic restraints requires an accurate analysis of structural behaviour when the actions increase from the serviceability level to the ultimate one. In the paper a probabilistic analysis, able to consider the scattering of the main parameters involved, is firstly proposed for such kind of problems. Then, by means of a sensitive analysis applied to the main parameters that demonstrate a significant influence on the structural behaviour, guidelines and design rules are proposed for the new generation of design codes.

Tailor Made Concrete Structures – Walraven & Stoelhorst (eds)
© 2008 Taylor & Francis Group, London, ISBN 978-0-415-47535-8

Concrete stress blocks of MC 90 and EC2. How safe are they?

L.C.D. Shehata
Universidade Federal Fluminense, Niterói, Brasil

A.L. Paula
Engevix, Rio de Janeiro, Brasil

I.A.E.M. Shehata
COPPE- Universidade Federal do Rio de Janeiro, Rio de Janeiro, Brasil

ABSTRACT: The utilization of High Strength Concrete (HSC) is more advantageous in structural elements under compression and the compressive stress-strain relationship is a relevant characteristic of the concrete needed for the design of those elements. For cross-sections design, the CEB-FIP MC 90 (1993) and the EN1992-1-1 (2004) permit curved-rectangular idealized concrete stress blocks and simplified rectangular ones. If the level of the axial load is relatively high, those stress blocks can lead to quite different strength capacity of HSC cross-sections. The experimental strengths of 403 elements subjected to pure axial load or to combined axial load and bending moment found in the literature are compared to the theoretical ones obtained considering those compressive stress diagrams for the concrete. The analysis of that comparison gives an idea about the level of safety related to the different design procedures, relevant information for codes writers and those who want to design HSC structures.

Tailor Made Concrete Structures – Walraven & Stoelhorst (eds)
© *2008 Taylor & Francis Group, London, ISBN 978-0-415-47535-8*

Shear design of FRC members with little or no conventional shear reinforcement

F. Minelli & G.A. Plizzari

University of Brescia, Department DICATA, Brescia, Italy

ABSTRACT: The present paper deals with some crucial design aspects of Fiber Reinforced Concrete (FRC) beams under shear loading, with or without conventional transverse reinforcement. It focuses on shear critical beams made of plain concrete or FRC. The influence of fibers on crack formation and development, failure mode, ductility and stiffness are herein investigated. A recent analytical proposal of the authors is further developed. Statements on the minimum shear reinforcement provided by steel fibers are derived and discussed. Finally, some preliminary results on the size effect issue of FRC members in shear and a design example are reported.

Tailor Made Concrete Structures – Walraven & Stoelhorst (eds)
© 2008 Taylor & Francis Group, London, ISBN 978-0-415-47535-8

A new future-oriented Model Code for concrete structures

J.C. Walraven
Delft University of Technology, Delft, The Netherlands

A.J. Bigaj - van Vliet
TNO Built Environment and Geosciences, Delft, The Netherlands

ABSTRACT: Codes have always played an important role in the design of structures. In the past, CEB and FIP devoted considerable attention to the actualisation of codes, especially by writing the Model Codes for Concrete Structures. Those codes were intensively used as a basis for updating national codes and as a source for providing a new European code – the Eurocode, which is nowadays accepted as hEN. However, code writing is still at the centre of interest. It is felt that new developments ask for new ideas about the basis and the content of codes in the future. In this paper some general reflections are given to the renovation of codes and progress in drafting the new *fib* Model Code for concrete structures.

A model for SFRC beams without shear reinforcement

P. Colajanni, A. Recupero & N. Spinella
University of Messina, Messina, Italy

ABSTRACT: In this paper a physical model, for the prediction of ultimate shear strength of Steel Fibers Reinforced Concrete (SFRC) beams is developed from the plastic Crack Sliding Model (CSM) introduced by Zhang (1997), based on the hypothesis that cracks can be transformed into yield lines. In this work the effectiveness factors are recalculated for SFRC beams and some further developments are introduced in the CSM, taking into account the fundamental post cracking tensile strength contribute of SFRC. The proposed model is validate by a large set of tests collected in literature and some numerical analyses were carried out to show the influence of fibers on the failure beams mode.

Modifying and adapting structures

Tailor Made Concrete Structures – Walraven & Stoelhorst (eds)
© 2008 Taylor & Francis Group, London, ISBN 978-0-415-47535-8

Seismic response assessment and upgrading of a complex of seven RC buildings using FRPs

T.B. Panagiotakos, A.J. Kosmopoulos & B. Kolias
DENCO Engineering Consultants S.A., Athens, Greece

ABSTRACT: A complex of seven reinforced concrete buildings constructed in the late 1970s to house the Data Processing Centre of the Bank of Greece was upgraded seismically in order to withstand (although marginally) the design earthquake with Peak Ground Acceleration (PGA) of 0.16 g prescribed by the current code. The evaluation of the buildings and the design of its retrofitting were realized according to Part 3 of Eurocode 8 (EN1998-3), published in 2005 as the European Standard for seismic assessment and retrofitting of buildings, including the use of externally bonded Fiber Reinforced Polymers (FRPs).

Tailor Made Concrete Structures – Walraven & Stoelhorst (eds)
© 2008 Taylor & Francis Group, London, ISBN 978-0-415-47535-8

First full scale application of a structure strengthened with organic prestressing – A case study

P. Pacheco, A. Guerra, P. Borges & H. Coelho
Faculty of Engineering & BERD SA (Bridge Engineering Research and Design), Porto, Portugal

ABSTRACT: The first full scale application of a movable scaffolding system strengthened with an Organic Prestressing System (OPS) is briefly described. The main characteristics of the steel structure and of the OPS technology are presented and significant aspects of the equipment's structural behaviour are given. The advantages of this innovative solution are established. Results prove that this control system enables the design of lighter scaffolding systems, reducing their service deflection and consequently making the construction easier and quicker.

Tailor Made Concrete Structures – Walraven & Stoelhorst (eds)
© 2008 Taylor & Francis Group, London, ISBN 978-0-415-47535-8

Eliminating bridge piers using stay cables with unconventional layouts

A.M. Ruiz-Teran
School of Computing and Technology, University of East London, London, UK

A.C. Aparicio
Department of Construction Engineering, Technical University of Catalonia, Barcelona, Spain

ABSTRACT: This communication explores innovative schemes and shows how in continuous prestressed concrete viaducts of medium length (40 m) the inclusion of under-deck or combined stay cables allows the elimination of intermediate or end piers. In these cases, the length of certain spans is doubled, while the main characteristics of the deck (depth, concrete strength, amount of reinforcement, amount of active steel, etc.) can be maintained. These schemes are very appropriate for the situations in which non-structural conditions make a uniform span distribution a non possible option for the design of a viaduct. In addition, it could be an alternative option when a pier of a built bridge has to be shifted due to the widening of an underneath infrastructure.

Tailor Made Concrete Structures – Walraven & Stoelhorst (eds)
© 2008 Taylor & Francis Group, London, ISBN 978-0-415-47535-8

The widening of Los Santos bridge. A case study of a tailor-made structure

H. Corres Peiretti, A. Pérez Caldentey, J. Romo, J. León González, F. Prieto &
J. Sánchez Delgado
FHECOR Ingenieros Consultores[1]

ABSTRACT: In this paper the main features of the Construction project of the widening of Los Santos Bridge are discussed. The project involves the widening of a major 600 m long bridge with spans of 150 meters increasing the width of the deck from 12.00 to 24.00 meters, without recurring to an independent structure. The project is based on the idea of providing minimum strengthening of the existing structure. This approach has allowed very important savings but has also required a challenging design project some aspects of which are discussed in this paper.

---

[1] In addition to the authors of this paper, in the different phases of development of this project other engineers working for FHECOR have participated. Very valuable information was also contributed by M. Martín Pardina, L.M. Viartola and L. Peset of DRAGADOS.

Tailor Made Concrete Structures – Walraven & Stoelhorst (eds)
© 2008 Taylor & Francis Group, London, ISBN 978-0-415-47535-8

Strengthening and design of shear beams

N. Randl
Carinthia University of Applied Sciences, Spittal, Austria

J. Kunz
Hilti Corp., Schaan, Liechtenstein

ABSTRACT: Concrete beams may fail in shear depending on kind of loading and amount of shear reinforcement. A research project has been started to investigate a new method of strengthening beams with insufficient shear resistance by applying post-installed reinforcement and in parallel derive an adequate model for shear design. Number and location of the inclined rebars as well as type of injection mortar has been varied. The rebars were installed in mortar-injected boreholes and anchored with metal plates at the accessible bar end. The test results confirmed that post-installed rebars can significantly increase the beam shear resistance provided they are situated properly and adequate injection mortars used. Above that the evaluation of the different contributions to shear resistance like truss action, dowel action and shear strength of compression chord provides new findings on the general shear failure mechanism in RC structures.

Tailor Made Concrete Structures – Walraven & Stoelhorst (eds)
© 2008 Taylor & Francis Group, London, ISBN 978-0-415-47535-8

Assessment of remaining structural capacity by computer simulation

J. Cervenka, V. Cervenka, R. Pukl & Z. Janda
Cervenka Consulting, Prague, Czech Republic

ABSTRACT: Further exploitation of existing concrete structures requires an assessment of the remaining structural capacity in the serviceability as well as ultimate limit states. This may be accomplished by computer simulation based on non-linear structural analysis. Such simulation exploits the state-of-the-art knowledge of material engineering, numerical solution methods and modern software technology. Commercial program ATENA developed by authors is suitable for this purpose. Examples of a railway bridge and a building slab structure are show.

Tailor Made Concrete Structures – Walraven & Stoelhorst (eds)
© 2008 Taylor & Francis Group, London, ISBN 978-0-415-47535-8

Experimental tests on repaired and retrofitted bridge piers

T. Albanesi, D. Lavorato, C. Nuti & S. Santini
Department of Structures, University of Roma Tre, Rome, Italy

ABSTRACT: This research aims to study the seismic performance of existing r.c. bridge piers specimens heavily damaged after previous pseudodynamic tests and actually repaired and upgraded by using self compacting concrete, stainless steel rebars and CFRP wrapped strips. Pier specimens are representative of tall and squat circular r.c. piers designed according to Eurocode 8 and Italian Code before 1986: EC8 columns are repaired and Italian ones retrofitted to enhance ductility and shear capacity. Innovative materials have been used and such interventions are described in details with particular attention to practical problems occurred. Experimental test results carried out at the *Laboratory of experiments on materials and structures* of the University of Roma Tre on self compacting concrete and stainless steel bars are shown and discussed. Retrofitting design and equipment for pseudodynamic tests are also presented. In coming tests aim to evaluate the effectiveness of adopted repairing and upgrading techniques not only to increase ductility but shear strength too.

Tailor Made Concrete Structures – Walraven & Stoelhorst (eds)
© 2008 Taylor & Francis Group, London, ISBN 978-0-415-47535-8

Enhanced safety with post-installed punching shear reinforcement

J. Kunz
Hilti Corp., Schaan, Liechtenstein

M. Fernández Ruiz & A. Muttoni
Ecole Polytechnique Fédérale de Lausanne, Lausanne, Switzerland

ABSTRACT: A considerable number of flat slabs supported by columns need to be strengthened against punching shear. Reasons are increasing loads, construction or design errors, but also more stringent code requirements due to the increased knowledge gained in the past years. This paper shows the effects of bonding post-installed shear reinforcement into inclined holes drilled from the bottom of the slab. Laboratory tests have shown that this method not only increases the slab strength but also adds significant deformation capacity. The evaluation of the tests has resulted in a clear design concept based on the critical shear crack theory. First practical implementations have proved that the system can be installed economically at a total cost which is below that of other strengthening methods.

Tailor Made Concrete Structures – Walraven & Stoelhorst (eds)
© 2008 Taylor & Francis Group, London, ISBN 978-0-415-47535-8

Strengthening of prestressed viaducts by means of a reinforced concrete overlay

R.W. Keesom
BAM Infraconsult, Gouda, The Netherlands

W.J. Bouwmeester – van den Bos
BAM Infraconsult, Gouda, The Netherlands,
University of Technology, Faculty of Civil Engineering and Geosciences, Delft, The Netherlands

M. van Kaam
BAM Infraconsult, Gouda, The Netherlands

A.Q.C. van der Horst
BAM Infraconsult, Gouda, The Netherlands,
University of Technology, Faculty of Civil Engineering and Geosciences, Delft, The Netherlands

ABSTRACT: Many viaducts in the Netherlands have been designed for traffic loads that are smaller and less intense than required by the present day standards. This means that for maintaining the structure's primary function, recalculation and often strengthening is required to guarantee safety and durability. For the viaducts in highway A9 near Amsterdam strengthening by a reinforced concrete overlay is used. This paper describes material testing of the existing structure, FEM-modeling of the existing and overlay structures and some aspects of the detailed design of the concrete overlay.

Tailor Made Concrete Structures – Walraven & Stoelhorst (eds)
© 2008 Taylor & Francis Group, London, ISBN 978-0-415-47535-8

Advanced numerical design for economical cathodic protection for concrete structures

R.B. Polder & W.H.A. Peelen
TNO Built Environment and Geosciences, Delft, The Netherlands

F. Lollini, E. Redaelli & L. Bertolini
PoliTecnico di Milano, Milano, Italy

ABSTRACT: Concrete structures under aggressive load may suffer chloride induced reinforcement corrosion, in particular with increasing age. Due to high monetary and societal cost (non-availability), replacement is often undesirable. Durable repair is necessary, e.g. by Cathodic Protection (CP). CP involves an electrical current through the concrete to the reinforcement from an external anode. The current causes steel polarisation, electro-chemical reactions and ion transport. CP systems are designed from experience, which results in conservative designs and their performance is a matter of wait-and-see. Using numerical models for current and polarisation distribution, CP systems can be designed for critical aspects and made more economical. This paper presents principles and results of preliminary numerical calculations for design of CP systems, applied to protection of local damage in bridges (e.g. at leaking joints).

Architectural concrete

Tailor Made Concrete Structures – Walraven & Stoelhorst (eds)
© 2008 Taylor & Francis Group, London, ISBN 978-0-415-47535-8

Precast facades elements for the new Museum of Acropolis in Athens

Athanasios N. Apergis
NTU Athens
PROET SA Member of JP-Avax Concern

ABSTRACT: A tendency of our times as regards the front parts of a building is to be made of precast façade elements with architectural concrete.

The new Museum of Acropolis in Athens, which will house the ancient sculptures of Acropolis, after an international competition was assigned to and designed by the architects B. Tschumi (Switzerland) and M. Fotiadis (Greece). The architects designed for two of the four floors a frontage that consists of precast façade elements made of reinforced concrete.

There are about 145 pieces of precast façade elements, with average surface dimensions $2,78 \times 4,78$ m and width 0,30 m. Total surface 1.674,61 m^2. Almost all the precast façade elements have the same external dimensions but different structures. All precast façade elements must have the same colour and the same external surface. The constructor must achieve a unified surface shade of the precast façade elements without having to paint or repair the façade elements.

Tailor Made Concrete Structures – Walraven & Stoelhorst (eds)
© 2008 Taylor & Francis Group, London, ISBN 978-0-415-47535-8

Thin post-tensioned concrete shell structures

S. Dallinger & J. Kollegger

Institute of Structural Engineering, Vienna University of Technology, Vienna, Austria

ABSTRACT: Shells are optimal structures. They can carry maximum loads over long spans on all scales, from nano-tubes to cooling towers with a very low amount of construction material. Today the use of shells is gaining more importance in architectural designs for facades interior design and roof structures. For the creation of large surface structures a viable connection technique is necessary. The connection of the elements will be established by post-tensioning. This allows for a blunt connection, that is able to transfer normal forces and moments across the connection. This connection technology was tested in experimental investigations. Finally, a feasibility study was carried out by erecting a very slender arch using this connection technology.

Tailor Made Concrete Structures – Walraven & Stoelhorst (eds)
© 2008 Taylor & Francis Group, London, ISBN 978-0-415-47535-8

Infra-lightweight concrete

M. Schlaich & M. El Zareef

Institute of Structural Engineering, Technische Universität, Berlin, Germany

ABSTRACT: Fair-faced concrete does not only possess high visual qualities. Monolithic concrete structures are also particularly durable, and the fact that no plastering or cladding is required leads to cost savings and makes buildings more sustainable and easier to recycle. However, due to the high thermal conductivity of normal concrete, fair-faced concrete without insulation causes prohibitive heating costs in cold countries. Infra-lightweight concrete with a dry bulk density of less than 800 kg/m^3 and the corresponding advantageous thermal properties promises to overcome this problem while maintaining the advantages. At the Technical University in Berlin such infra-lightweight concrete was developed and to prove its practicality a single family house was built with it using glass fiber bars as reinforcement. This paper describes this new concrete mix and its properties. It elaborates on the structural implications when working with infra-lightweight concrete. Design and construction of the house will also be presented.

Tailor Made Concrete Structures – Walraven & Stoelhorst (eds)
© 2008 Taylor & Francis Group, London, ISBN 978-0-415-47535-8

Challenging concrete structure with a blend of architectural fair faced concrete

Vinay Gupta
Tandon Consultants Pvt Ltd, New Delhi, India

ABSTRACT: To celebrate 500 years of history of Sikhs and 300 years of establishment of the Khalsa, the mega project 'Khalsa Heritage Complex' was launched by the Punjab Government. The project comprises a 150m long pedestrian bridge to connect Complex 'A' and Complex 'C'. While the Complex 'A' houses library blocks and a theatre, the Complex 'C' mainly houses high-tech exhibits of various types. The special highlights of the project include (i) 26 m span prestressed concrete ramps in Heritage Museum building (ii) 20 m span RCC roof beams acting as partial catenary in Permanent Exhibit building (iii) 35 m span arch bridges incorporating prestressed tie beams (iv) Precast canopy over the pedestrian bridge, (v) Specialized Mechanical Connection between in-fill brick walls and the adjoining beam-column frame structure for sustainability during high seismic forces (vi) Inception of large volume of architectural fair faced concrete (vii) Preparation of mock ups of all specialized elements, prior to their actual construction.

Tailor Made Concrete Structures – Walraven & Stoelhorst (eds)
© 2008 Taylor & Francis Group, London, ISBN 978-0-415-47535-8

Bond behaviour between GFR bars and infra-lightweight concrete

M. El Zareef & M. Schlaich
Institute of Structural Engineering, Technische Universität, Berlin, Germany

ABSTRACT: The purpose of this paper is to investigate the bond behaviour between Infra-lightweight Concrete (ILWC) and different types of reinforcement such as Glass-Fibre Reinforcement (GFR) and steel reinforcement (RFT). Another objective is to improve the bond-slip relationship in infra-lightweight concrete by enhancing the tensile strength of infra-lightweight concrete using polypropylene (PP) fibres with various lengths or using different confinement ratios. For these objectives 27 specimens were tested for bond stress and slip between ILWC and different bars. A special mould was produced to ensure the rebar position during casting. The bar was pulled out with a loading rate of 0.005 mm/s. The slip was measured from the dead end of the bar. The study concludes that the configuration of the ribs on the surface of bars plays the main role on the bond-slip relation especially at the same concrete strength. Improving of radial tensile strength by using confinement stirrups or by adding PP fibres with length up to 20 mm in ILWC improves the bond behaviour especially with bars that have more ribs per unit length. However, adding of PP fibres reduced the slip between bars and ILWC at the maximum bond stress, which is required to control the cracks width in serviceability limit state.

Tailor Made Concrete Structures – Walraven & Stoelhorst (eds)
© 2008 Taylor & Francis Group, London, ISBN 978-0-415-47535-8

Design and construction of open deck bridge

T. Abo, M. Ooba, S. Yoda & S. Suzuki
East Japan Railway Company, Tokyo, Japan

ABSTRACT: A railway bridge intersecting a river was to be reconstructed, as part of a river improvement project. It was necessary to reconstruct the bridge at a length of 119.2 meters, as the old bridge was a steel deck bridge of 52.78 meters, and obstructed flood flow. The most effective way to keep construction cost at a minimum is to limit the roadbed height at access. Therefore, in selecting the style of the new fin-back type bridge, through-bridge form was applied to keep the rail level as low as possible, and in consideration of aesthetics, at three-span continuation PRC through-bridge was chosen. In addition, the bridge is an open deck type which has openings in the deck as the bridge is located in one of the heaviest areas of snowfall in Japan, so it was necessary to take measures to cope with snowfall.

Tailor Made Concrete Structures – Walraven & Stoelhorst (eds)
© 2008 Taylor & Francis Group, London, ISBN 978-0-415-47535-8

New Amsterdam Public Library (Openbare Bibliotheek Amsterdam)

Joop Paul, Frank van Berge Hengouwen, Hugo Mulder & C. Tait van der Putten
Arup Amsterdam, The Netherlands

ABSTRACT: The new Amsterdam Library is the largest library in Europe. Arup provided structural, MEP and lighting design services. Through an integrated approach to architecture and engineering it was possible to create a stunning building. The building is equipped with energy saving measures such as long term energy storage and the use of the structural frame as thermal mass. The plenums underneath the raised floors and the large vertical atrium are used to distribute and extract air. By integrating structural and MEP elements, height could be reduced, making it possible to add one extra floor within the building envelope. A carefully designed grid, compatible with both the below ground car park and the library building, and the theater box, which is situated on the top floors of the building, prevented the need for transfer structures. This allowed for a further reduction in height. By placing cores outside the building, the use of spectacular sculptural columns and suspending structural elements such as walls and columns, an exciting dynamic design was achieved.

Tailor Made Concrete Structures – Walraven & Stoelhorst (eds)
© 2008 Taylor & Francis Group, London, ISBN 978-0-415-47535-8

Concrete in the optimal network arch

Per Tveit

Agder University, Grimstad, Norway

ABSTRACT: The network arch is an arch bridge with inclined hangers that cross each other at least twice. If hangers cross each other only once, it is a Nielsen bridge. When the ties are a thin concrete slab, the network arch uses less than 2/3 of the steel needed for arch bridges with vertical hangers. A network arch is likely to remain the world's most slender arch bridge. The arches can be made from universal columns of American wide flange beams that come from the steel works with the desired curvature. In many equal spans in long bridges the arches can be made from high strength concrete. Network arches have a pleasing appearance and do not block the view to the landscape and cityscape behind them. A temporary tie can be used in the erection. Spans of up to 300 m can be finished on shore and be lifted onto the pillars by big floating cranes.

Tailor Made Concrete Structures – Walraven & Stoelhorst (eds)
© 2008 Taylor & Francis Group, London, ISBN 978-0-415-47535-8

The new Lisbon indoor sports complex

J.N. Bastos
Faculdade de Arquitectura- Universidade Técnica de Lisboa, Lisbon, Portugal

ABSTRACT: The recently inaugurated Lisbon Indoor Sports Facility represents one of the major achievements of the City of Lisbon regarding public sports equipments. The 15,000 sq. mtr. Sports Complex facilities consist of several building units, particularly, a 50.0 m × 50.0 m indoor sports pavilion with 1,000 seats. A 300 car spaces underground parking area was required. The materials being used were reinforced concrete (RC) for the general structural system and structural steel lattice space trusses for the roof systems.

The purpose of this study is to present the difficulties created by the need of designing large span roof structures that are simultaneously aesthetically pleasant and economically sound, and integrate them with the extensive RC structural system. Several design solutions were used and an useful design comparison can be made relating the main advantages/shortcomings of each solution.

Tailor Made Concrete Structures – Walraven & Stoelhorst (eds)
© 2008 Taylor & Francis Group, London, ISBN 978-0-415-47535-8

New Vodafone building in Oporto – A white concrete jagged shell

C.M. Quinaz & A.P. Braga
Afaconsult, V. N. Gaia, Portugal

ABSTRACT: The new Vodafone building in Oporto is a five storey high, with three basement levels, openspace office building. It will certainly become an architectural landmark for the city. The 7336 m^2 slab building is defined by its highly irregular white concrete façade. The purpose of this irregularity is to convey a "sensation" of "motion", as this is the brand image of the Client (Vodafone). The development of the geometry and the structural design was made by Afaconsult (Portugal).

Tailor Made Concrete Structures – Walraven & Stoelhorst (eds)
© 2008 Taylor & Francis Group, London, ISBN 978-0-415-47535-8

Innovative footbridges used as urban furniture for our cities for the future

A. González Serrano

Proxectos SL. Coruña, Spain

ABSTRACT: This publication shows several singular urban footbridges that I've designed as structural elements integrated as part of an urban street furniture, composed by light, slender and elegant designs. They are cable-stayed constructions made of concrete and steel, with metallic decks or concrete slabs. The pavement is in wood or concrete. Attractive and distinct, they are conceived to provide continuity to pedestrian routes, or joining areas with heavy traffic, also having the function to define spaces, thus creating distinct landmarks.

Tailor Made Concrete Structures – Walraven & Stoelhorst (eds)
© 2008 Taylor & Francis Group, London, ISBN 978-0-415-47535-8

Fabric formwork for flexible, architectural concrete

N. Cauberg & B. Parmentier
Belgian Building Research Institute, Brussels, Belgium

D. Janssen
Centexbel, Ghent, Belgium

M. Mollaert
Vrije Universiteit Brussel, Brussels, Belgium

ABSTRACT: An innovative research project scanned the possibilities to use fabric as a flexible formwork for architectural concrete elements. Besides a general feasibility study, the project focuses on the textile parameters such as stiffness and permeability, the quality of the concrete surface and the modeling of the formwork both before and after casting. For this, the project gathers together textile industry, architects and contractors, to combine architectural creativity and modern technology. The first laboratory tests show the potential of the concept, and a range of different shapes of formwork and concrete elements. A series of case studies has been elaborated to deal with issues as textile choice, shape modeling, textile pretension, formwork fixation, concrete application and more. This article presents a general outline of fabric formworks and the details for some case studies such as architectural columns and double curved shells.

Developing a modern infrastructure

Tailor Made Concrete Structures – Walraven & Stoelhorst (eds)
© 2008 Taylor & Francis Group, London, ISBN 978-0-415-47535-8

Offshore foundation in concrete – Cost reduction by serial production

Hugo Mathis
RSB Schalungstechnik GmbH & Co, Fussach, Austria

Tailor Made Concrete Structures – Walraven & Stoelhorst (eds)
© 2008 Taylor & Francis Group, London, ISBN 978-0-415-47535-8

Design and construction of an immersed concrete tunnel using an integrated dock facility

C. Bauduin
BESIX, Brussels, Belgium
University of Brussels, Brussels, Belgium

P. Depuydt
BESIX, Brussels, Belgium

ABSTRACT: The construction of the A73 highway (the Netherlands) involved the construction of a 2.4 km long, 2*2 lane tunnel to mitigate the impact of traffic on the city of Roermond and to cross the 1 km wide valley of the river Roer. Given the ground conditions (mainly dense sand and gravel), required cross section and tunnel depth, an immersed tunnel appeared to be the most economic solution to cross the valley. However, a facility to permit precasting of the RC tunnel elements needed to be established. The small depth and width of the Roer prevented transportation of tunnel elements by use of the river and dewatering was not permitted due to the associated environmental impact. The analysis of the geotechnical data indicated the presence of a 5 m thick local loam layer over a length of 350 m along the tunnel alignment, located at the eastern part of the Roer valley. This impervious layer offered the opportunity to excavate a 350 m long dock along the axis of the tunnel between temporary anchored sheetpiles that were installed into the loam layer. Two precast tunnel elements of approximately 158 m long could be constructed in this dock. A trench was excavated between and temporary sheetpiles were installed 5 m outside the future tunnel location, thus permitting the transportation and immersion of a total of four such tunnel elements in two installments from the dock to their final location. The remaining part of the tunnel was constructed in what was previously the dry dock after the immersion of the elements. This paper describes how the concrete structure was designed and specified (concrete weight and tolerances) with allowances for the specific geotechnical and hydraulic conditions (water depth, concrete weight and freeboard, water level management in the trench, excavation depth and uplift of the impervious layer etc.). The paper describes the behavior and provisions of the concrete structure on the gravel bed foundation, which was preferred to sand flow to minimize the risks of liquefaction as the area is mode-rately seismic.

Tailor Made Concrete Structures – Walraven & Stoelhorst (eds)
© 2008 Taylor & Francis Group, London, ISBN 978-0-415-47535-8

Bridges over Timbabé river: High performance concrete in the middle of the jungle

B.P. Van den Bossche
Concrete specialist – Besix group

ABSTRACT: Equatorial Guinea is one of the small countries at the west coast of Africa. The existing road and harbour infrastructure are old and not suitable for the growing market. To meet with the grow of the capital Malabo, on the isle of Bioko, a double lane ring way through the Jungle is built. The two bridges over the Timbabé River will be the biggest arc construction on the Ilse.

The bridges are made by a post-tensioned cantilevered concrete construction, with a joint in the middle of the span. The cantilevers are supported by a mass construction; which is supported on the natural basaltic rock.

This paper will give an overview on the preparations of the concrete mix, the problems encountered caused by the local conditions and the quality follow up of the concrete.

Tailor Made Concrete Structures – Walraven & Stoelhorst (eds)
© 2008 Taylor & Francis Group, London, ISBN 978-0-415-47535-8

Precast segmental design and construction in China

Dong Xu, Huichi Li & Chao Liu

Dept. of Bridge Engineering, Tongji University, Shanghai, China

ABSTRACT: In 2001, Liuhe River Bridge in Shanghai was the first span-by-span precast segmental bridge constructed by a launching gantry in China. In 2003, Humin Viaduct in Shanghai was the first project using short line segment casting for urban elevated viaducts. The biggest PC beam bridge using balanced cantilever precast segmental construction in China is Jiujiang Bridge in Guangdong province, which has two 160 m main spans and was completed in 1996. In July 2001, the first 4-span cable-stayed bridge in China, Yiling Bridge in Hubei province, using precast segmental PC box girder construction, was completed. Although China has these successful references, the precast segmental construction, originated from the contractor system, was not well followed in time by design codes. Lack of specified design codes in China usually causes owners and even designers hesitate to use this construction method. Longitudinal mild reinforcing bars do not continue through the joints. The new concept of making closed horizontal shear reinforcement, with the addition of normally arranged vertical stirrups, to carry the horizontal and vertical components of the diagonal tensile stress was proposed in the reinforcement design of precast segments for the approach bridges of the Sutong Bridge.

Tailor Made Concrete Structures – Walraven & Stoelhorst (eds)
© 2008 Taylor & Francis Group, London, ISBN 978-0-415-47535-8

The use of robots and self-compacting concrete for unique concrete structures

L.N. Thrane, T.J. Andersen & D. Mathiesen
Danish Technological Institute, Taastrup, Denmark

ABSTRACT: Today's concrete architecture is often dominated by repetitiveness and recognisable geometries like squares and rectangles. Digitally designed and extraordinary concrete architecture involves excessively high construction costs, due to the production methods based on craftsmanship. A way of opening up the prospect of a new and exciting concrete architecture is to find new automated industrial methods for the production of singular concrete structures. The aim of an ongoing Danish project called "Unique Concrete Structures" is to build alternative moulds by using robots and tailor-made self-compacting concrete that will spread in the moulds and thus shape concrete structures according to the architect's instructions. A new High Technology Concrete Laboratory equipped with a robot cell and a fully automatic mixing plant has been established at the Danish Technological Institute.

Tailor Made Concrete Structures – Walraven & Stoelhorst (eds)
© 2008 Taylor & Francis Group, London, ISBN 978-0-415-47535-8

Behavior of a multiple spans cable-stayed bridge

S. Arnaud, N. Matsunaga, S. Nagano & J.-P. Ragaru
Ingerosec Corp., Tokyo, Japan

ABSTRACT: Bridges with multiple cable-stayed spans are more and more designed for large crossing projects, but the behavior of such bridges is by far different from the behavior of a standard cable-stayed bridge, with additional problems of stiffness in the middle spans and thermal expansion of the deck. We got the opportunity to participate in the design check of a five towers cable-stayed bridge with 300 meters spans and we examined the configuration between type of connection, stiffness of deck, stiffness of piers and pylons, in order to confirm the minimal structural cost. We analyzed on a FEM model the differences in the forces distribution following the connection case and we organized further calculations about the relationship between stiffness of deck, pylons and piers. Results are presented with particular focus about the impacts of asymmetric loading and thermal expansion of the deck on this multiple spans structure.

Tailor Made Concrete Structures – Walraven & Stoelhorst (eds)
© 2008 Taylor & Francis Group, London, ISBN 978-0-415-47535-8

Development of a new viaduct structure to achieve a high-amenity under-viaduct space

Hisahide Sugisaki, Kaoru Kobayashi, Taichiro Watanabe & Seiji Ikeno
East Japan Railway Company, Frontier Service Development Laboratory, Saitama, Japan

ABSTRACT: This paper explains a new viaduct structure that improves the environment of under-viaduct space. We thought out a new structure for a low-vibration viaduct to reduce the vibration transmitted from beams to columns by inserting elastic material to the joints of beams and columns. With this socket joint structure with sufficient socket joint depth, it is possible to give columns the load carrying capacity equivalent to columns of traditional structures. We checked the low-vibration effect by analyzing trains running in the viaduct with socket joints and confirmed that the vibration acceleration level in the 30–60 Hz band can be reduced by approximate 20–25 dB compared to traditional structures. In order to check a seismic performance, we carried out reversal cyclic loading tests, and analyzed seismic response. This research proved that our new structure has the effect of reducing damage of columns, while seismic deformation of upper points is larger than traditional structures.

Tailor Made Concrete Structures – Walraven & Stoelhorst (eds)
© *2008 Taylor & Francis Group, London, ISBN 978-0-415-47535-8*

Study on structural behavior characteristics of concrete filled steel tube girder bridges

Won Jong Chin, Jae Yoon Kang, Eun Suk Choi & Jung Woo Lee
Korea Institute of Construction Technology, Goyang, Republic of Korea

ABSTRACT: The new bridge system described in this paper uses concrete-filled steel tube girders instead of conventional girders. A new type of shear connector attached to the tube that is ¬-shaped perfobond rib provides the means for developing composite action between the steel tube and the concrete slab deck. Experimental investigations were carried out to examine the flexural behavior of the proposed concrete filled steel tube girder. The experimental investigation consisted of designing and constructing a test specimen and loading it to collapse in bending to understand the ultimate capacity of the system. Test results showed that concrete filled steel tube girder has good ductility and maintains its strength up to the end of the loading. Results of this investigation demonstrated the potential for using a concrete filled tube as bridge girder. Additional experimental investigations were carried out to examine the fatigue behavior of the concrete-filled steel tube girder and revealed also that the new type of girder bridge exhibits good fatigue durability.

Tailor Made Concrete Structures – Walraven & Stoelhorst (eds)
© 2008 Taylor & Francis Group, London, ISBN 978-0-415-47535-8

Immersed parking facilities

R.J. van Beek & H.M. Vlijm
Witteveen + Bos Consulting Engineers

ABSTRACT: The development of parking facilities in urban areas is a difficult task. Space is scarce and parked cars on surface level are an eyesore in historical cities. The construction of parking facilities above ground level or underground occupies rare space and creates nuisance during the construction activities and is costly. Witteveen + Bos developed a solution for saving of scarce space and a new construction method for car parks. This method can be applied in a great number of the 5,000 (sea) ports in the world. The solution is the construction of car parks below the water table, in old harbour basins by the application of the immersed tunnelling method. The immersion method has been applied world wide for the construction of submergible traffic tunnels, In the Netherlands 30 immersed tunnelling projects were successfully completed. Witteveen + Bos, through its partnership "Tunnel Engineering Consultants" was involved with many of these tunnelling projects. To our knowledge no project was realised yet with the application of the immersion of building elements for the construction of under water parking or storage facilities. Interesting locations are Amsterdam, Rotterdam, London, Marseille, Monaco and New York.

Tailor Made Concrete Structures – Walraven & Stoelhorst (eds)
© 2008 Taylor & Francis Group, London, ISBN 978-0-415-47535-8

Slipforming of advanced concrete structures

K.T. Fossa
Aker Kvaerner, Oslo, Norway

A. Kreiner
Gleitbau, Salzburg, Austria

J. Moksnes
Jan Moksnes Consulting, Stavanger, Norway

ABSTRACT: Slipforming is a method of constructing tall concrete structures based on known parameters and proven technology. The method encompasses several activities and the successful execution of a slipform operation depends on proper understanding of the mechanisms involved, careful planning and work preparation and the skills of the operator. Slipform operations are often executed by specialist subcontractors. Recent advances in concrete technology, pump line equipment and formwork design enable the method to be employed in the rapid erection of tall and demanding concrete structures, but schedule and quality advantages can also be achieved for structures of more ordinary size and complexity. The paper describes the important parameters for successful slipforming and contains some examples of slipforming of advanced concrete structures.

Tailor Made Concrete Structures – Walraven & Stoelhorst (eds)
© 2008 Taylor & Francis Group, London, ISBN 978-0-415-47535-8

Tagus Crossing at Carregado (Portugal): A project respectful of its sensitive environment

A. Perry da Câmara
Perry da Câmara e Assoc. Cons. Eng., Lisbon, Portugal

Alexandre Portugal
COBA, Lisbon, Portugal

Francisco Virtuoso
Civilser, Lisbon, Portugal

Michel Moussard
Arcadis, Le Plessis-Robinson, France

ABSTRACT: The Tagus Crossing at Carregado in Portugal is located in the Tagus river alluvial valley, which is one of the most fertile Portuguese agricultural areas. The crossing is approximately 11,700 m long, carrying a double 3 lane carriageway with 30 m of total width. It is made of three main structures: the 1700 m long North Viaduct, the 970 m long Tagus Bridge and the 9200 m long South Viaduct. Additionally to these main structures the environmental sensitivity of the agricultural ground occupation imposed the construction of a set of significant complementary works: an 8 m wide 9200 m long side road along the South Viaduct to serve the construction site, including 18 pontoons to overcome local waterlines, dikes and irrigation existing infrastructures. In this paper a description of the project is made with emphasis in the design and construction aspects associated with the limitation of impacts on the adjacent sensitive environment.

Tailor Made Concrete Structures – Walraven & Stoelhorst (eds)
© 2008 Taylor & Francis Group, London, ISBN 978-0-415-47535-8

High Rise Buildings. The challenge of a new field of possibilities for the use of structural concrete

H. Corres, J. Romo & E. Romero
FHECOR Ingenieros Consultores

ABSTRACT: In this paper, the design and construction of several buildings of moderate height (no more that 250 m) are analyzed. In all these projects structural concrete has been used for different elements: floorings, special steel-concrete composite columns using high performance concrete, Shear walls, stiffening floors, etc.

Tailor Made Concrete Structures – Walraven & Stoelhorst (eds)
© 2008 Taylor & Francis Group, London, ISBN 978-0-415-47535-8

Direct load transmission in sandwich slabs with lightweight concrete core

E. Schaumann, T. Vallée & T. Keller
Ecole Polytechnique Fédérale de Lausanne, Composite Construction Laboratory, EPFL-CCLab,
Lausanne, Switzerland

ABSTRACT: This paper presents a variational energy based model to predict the cracking and ultimate load of hybrid FRP-concrete sandwich bridge slabs under direct load transmission in the support region. The slab consists of three layers: a glass fiber-reinforced polymer composite (GFRP) sheet with T-upstands for the bottom skin, lightweight concrete (LC) for the core and a thin layer of ultra high performance reinforced concrete (UHPFRC) as top skin. Different LC types were used: a low and a high density sand lightweight aggregate concrete (SLWAC) and an all-lightweight aggregate concrete (ALWAC). A bottle-shaped compressive strut and a continuous transverse tensile tie allowed for stress redistribution after cracking due to LC softening. Experimental cracking and ultimate loads of deep short span beams could be accurately modeled in this way. The arch rise of the compressive strut decreased significantly after concrete cracking.

Tailor Made Concrete Structures – Walraven & Stoelhorst (eds)
© 2008 Taylor & Francis Group, London, ISBN 978-0-415-47535-8

The Metrolink Finback Bridge, Manchester

S.W. Jones
Gifford, Chester, UK

R.G. Wrigley
Gifford, Southampton, UK

ABSTRACT: The Manchester to Oldham Line of the proposed Metrolink Phase 3 tram system passes through the new Central Park Business Park utilising the existing track bed of a disused and dismantled heavy rail route. To the west of Thorp Road Bridge the new Metrolink route crosses the existing four tracks of the Manchester to Leeds Trans-Pennine railway. A post-tensioned concrete finback bridge has been provided to carry the new twin Metrolink tracks over the existing heavy rail route. In order to minimise disruption to the operational railway, the Finback Bridge was constructed in a position alongside to and roughly parallel to the existing railway. Once the construction was substantially complete, the 6,250 tonne bridge was rotated in plan through 21° to its final position during a weekend possession.

Tailor Made Concrete Structures – Walraven & Stoelhorst (eds)
© 2008 Taylor & Francis Group, London, ISBN 978-0-415-47535-8

Total precast solution for large stadium projects meet tight schedule

T.J. D'Arcy
The Consulting Engineers Group, Inc., San Antonio, TX, USA

ABSTRACT: A report on the successful application of precast prestressed concrete in large stadium projects. Framing system design and construction methods are presented and details of the construction and design of three major league and college stadiums are presented.

Tailor Made Concrete Structures – Walraven & Stoelhorst (eds)
© *2008 Taylor & Francis Group, London, ISBN 978-0-415-47535-8*

HPFRC plates for ground anchors

M. di Prisco, D. Dozio, A. Galli & S. Lapolla
Department of Structural Engineering, Politecnico di Milano, Milano, Italy

M. Alba
Surveying Department (DIIAR), Politecnico di Milano, Milano, Italy

ABSTRACT: In order to stabilize a ground slope, special plates of reduced sizes made of High Performance concrete reinforced with straight steel fibres were designed and built. The plates are reinforced also with special high bond steel bars with special steel threaded bushes welded at the ends to guarantee their tensile action on overall the slab size. Their weight is limited in order to assure the transportability by helicopter everywhere in mountain regions. After the experimental characterization of the material aimed to identify a constitutive relationship, a limit design approach was carried out. Ten plates were placed in situ and two of them were instrumented in order to follow the real stress state inside of the anchor plates: the main results are here described.

Designing structures against extreme loads

Tailor Made Concrete Structures – Walraven & Stoelhorst (eds)
© 2008 Taylor & Francis Group, London, ISBN 978-0-415-47535-8

Outer concrete containments of LNG-tanks – Design against thermal shock

Josef Roetzer
Dywidag International, Munich, Germany

Theodor Baumann
Munich, Germany

ABSTRACT: The stability and tightness of the concrete outer tank has to be guaranteed by an appropriate design also under cryogenic conditions. The relevant codes provide a lot of formal regulations, but give no precise indications for analytical procedures and criteria which have to be applied in the design. Hence, the following paper deals with the behaviour of reinforced and prestressed concrete sections in direct contact with LNG, considering thermal strains and consecutive crack formation. Mechanical models, which have to be clear and simple, are discussed. The course of sectional forces and displacements due to a temperature gradient of 180°C after failure of the inner tank are outlined for areas below and above the LNG level. Substantial design criteria are proposed and discussed. Essential in this respect are the thickness of the residual compressive zone, the reinforcement steel stresses and the characteristic crack width. By means of ingenious models, a way for the direct understanding of the coherence between the strains and deformations imposed by the temperature gradient and the above design criteria is pointed out.

Tailor Made Concrete Structures – Walraven & Stoelhorst (eds)
© 2008 Taylor & Francis Group, London, ISBN 978-0-415-47535-8

Economic aspects in design of enhanced earthquake resistant r/c buildings

M. Mezzi

Dip.Ingegneria Civile ed Ambientale, Università degli Studi di Perugia, Perugia, Italy

ABSTRACT: The direct cost comparison determines a wrong evaluation of the effectiveness of innovative seismic protection system, not accounting for the differences in the performance levels characterising the conventional and enhanced solutions. Deterministic and probabilistic procedures for the estimate of consequence parameters are evoked. Comparative analyses of the seismic response, in terms of typical demand parameters (story drift and floor acceleration) of conventional fixed-base and innovative base-isolated buildings are carried out. The evaluations account for the probabilistic nature of the external and internal influence parameters controlling the structure response that are managed with the statistics analytical tools. The main scattering source is the seismic input, but its consequence is practically zeroed by the isolation system performance. The other uncertainty sources, building characteristics, entity of the undergone damage, consequence of the damage, become irrelevant on the consequence evaluation on the occupants, costs and downtime.

Tailor Made Concrete Structures – Walraven & Stoelhorst (eds)
© 2008 Taylor & Francis Group, London, ISBN 978-0-415-47535-8

Seismic response of bridges on pile foundations considering soil-structure interaction

F. Dezi
Department of Materials and Environment Engineering and Physics, Università Politecnica delle Marche, Ancona, Italy

S. Carbonari
Department of Architecture, Constructions, Structures, Università Politecnica delle Marche, Ancona, Italy

G. Leoni
Department ProCAm, Università di Camerino, Ascoli Piceno, Italy

ABSTRACT: This paper attempts to assess the influence of dynamic soil-structure interaction on the behaviour of bridges founded on rigidly capped floating vertical pile groups. A numerical model for the analysis of the soil-structure interaction of generic structures on pile foundations is presented based on a finite element approach for superstructure and pile group whereas the soil is assumed to be a Winkler-type medium. The method is applied to single piers representative for a class of bridges. Varying the soil condition and the foundation geometry, same comparisons are made with respect to the fixed base model. Special issues such as the contribution of the soil profile, of the local amplification and of the rocking at the foundation level are discussed. Soil-structure interaction is found to be essential for effective design of bridges especially for squat piers and soft soil.

Tailor Made Concrete Structures – Walraven & Stoelhorst (eds)
© 2008 Taylor & Francis Group, London, ISBN 978-0-415-47535-8

Fastening technique in seismic areas: A critical review

C. Nuti & S. Santini

Department of Structures, University of Roma Tre, Rome, Italy

ABSTRACT: In the present paper, the authors focus their attention on the seismic capacity of anchors that transmit structural and non-structural loads from attachments into concrete members. In Europe, a large number of buildings are in seismic areas. As a consequence, anchors must be also qualified and designed for seismic loading according to european design criteria for buildings and other structures. Up to date provisions for qualification of products and design for application of fastener are still in progress. This work aims to clarify the understanding of the fundamental principles of the seismic behaviour. To this aim, the following crucial points are discussed and reviewed: cracking of concrete, requirements for ductility, overstrength and failure modes.

Tailor Made Concrete Structures – Walraven & Stoelhorst (eds)
© 2008 Taylor & Francis Group, London, ISBN 978-0-415-47535-8

MRI validation of FEM models to describe moisture induced spalling of concrete

S.J.F. Erich, A.B.M. van Overbeek & A.H.J.M. Vervuurt
TNO Built Environment and Geosciences, Delft, The Netherlands

G.H.A. v.d. Heijden, L. Pel & H.P. Huinink
Eindhoven University of Technology, Eindhoven, The Netherlands

ABSTRACT: Fire safety of buildings and structures is an important issue, and has a great impact on human life and economy. One of the processes negatively affecting the strength of a concrete building or structure during fire is spalling. Many examples exist in which spalling of concrete during fire has caused severe damage to structures, such as during the fires in the Mont Blanc and Channel Tunnel. Especially newly developed types of concrete such as HPC and SCC, have shown to be sensitive to spalling, hampering the application of these new concrete types. To reduce risks and building costs, the processes behind spalling need to be understood. Increasing our knowledge allows us to predict and take effective and cost friendly preventive measures. One of the mechanisms that drives spalling of concrete, is the heating of the moisture present inside concrete. When concrete is heated water will evaporate, which results in a high gas pressure inside the pores of the concrete.

In recent years much research has been performed on the processes behind spalling. Using this knowledge, a Finite Element Model (FEM) describing the moisture transport processes in heated concrete has been developed. However, the validity of all current models (including our own) is unknown because of debatable input parameters and lack of experimental techniques to follow the transport process in situ. In cooperation with the Eindhoven University of Technology moisture transport in heated concrete can now be investigated using a home built dedicated 1D Magnetic Resonance Imaging (MRI) setup. Using the results of the MRI experiments the validity of our FEM model is being assessed.

Tailor Made Concrete Structures – Walraven & Stoelhorst (eds)
© 2008 Taylor & Francis Group, London, ISBN 978-0-415-47535-8

Experience and tests of the fire-resistant plaster coating in bored tunnels

F.W.J. van de Linde
Nebest B.V., Groot-Ammers, The Netherlands

B.J. van der Woerd
CBBN Fireproofing Int./BAM Betontechnieken, Schiedam, The Netherlands

L. Mulder
CBBN Fireproofing Int./Vogel, Zwijndrecht, The Netherlands

ABSTRACT: CBBN Fireproofing Int., in which Vogel B.V. and BAM Betontechnieken are partners, has employed an innovative approach to apply a fire-resistant plaster to the two largest bored tunnels and a city tunnel in The Netherlands. The three main components of the method are a scientific test method, an optimized plaster and a robotized application technique. The method's success has been demonstrated in the Westerschelde Tunnel, the Groene Hart Tunnel and the Hubertus Tunnel. The method guarantees that the underlying concrete will not be damaged in the event of an extreme load such as a tunnel fire. With the right fire protection plaster the concrete has full protection and it is not spalling. The quality of the concrete has much influence on the design of the fire protection. Therefore the fire protection plaster is tested by Efectis. This paper describes the influence of the concrete on the fire-resistant material and will discuss test results and application of the fire-resistant plaster Fendolite MII.

Tailor Made Concrete Structures – Walraven & Stoelhorst (eds)
© *2008 Taylor & Francis Group, London, ISBN 978-0-415-47535-8*

The merits of concrete structures for oil and gas fields in hostile marine environments

T.O. Olsen
Dr.techn. Olav Olsen, Oslo, Norway

S. Helland
Skanska, Oslo, Norway

J. Moksnes
Jan Moksnes Consulting, Stavanger, Norway

ABSTRACT: Work is nearing completion in *fib* to produce a "State of the art" report on the performance of concrete offshore structures in hostile marine environments. The object of the report is to make available the experience gained over the past decades in the design and construction of such structures and to demonstrate the performance that can be achieved when attention is paid to known standards of quality and workmanship.

Concrete structures are now being considered for field developments in several parts of the world and in particular in Russia, where two concrete GBS's were installed in the Sakhalin Sea in 2005, and more are expected to follow, also in the arctic regions further north. The international concrete industry needs to position itself for new contracts in these regions where the conditions are tough and the competition is stiff from a strong steel lobby and a conservative mind set.

Tailor Made Concrete Structures – Walraven & Stoelhorst (eds)
© 2008 Taylor & Francis Group, London, ISBN 978-0-415-47535-8

MPU Heavy Lifter – A lightweight concrete vessel for heavy offshore lifting operations

T. Landbø
MPU Offshore Lift ASA

E.B. Holm & H. Ludescher
Olav Olsen a.s, Oslo, Norway

ABSTRACT: The paper describes concept and general design of a reinforced concrete vessel which has been developed for heavy offshore lifting operations. The vessel has overall dimensions of about 110 m × 87 m × 55 m and is able to lift up to 25 000 tonnes. It is currently under construction in Rotterdam and will be commissioned in the first quarter of 2009. Its key features are pivoting lifting frames and large flushing tanks that enable to lift entire topsides or jackets. Draft is controlled in a large scale by ballasting compartments of the concrete hull. For its propulsion, the vessel has 8 thrusters which are mostly required for positioning during lifting. For transportation the vessel will be towed by tug boats. Operations are controlled by a permanent crew that is accommodated and supplied by a living quarter with helicopter deck.

Tailor Made Concrete Structures – Walraven & Stoelhorst (eds)
© 2008 Taylor & Francis Group, London, ISBN 978-0-415-47535-8

Detail design of the MPU Heavy Lifter

H. Ludescher & S.A. Haugerud
Olav Olsen as, Oslo, Norway

M. Fernández Ruiz
Ecole Polytechnique Fédérale de Lausanne

ABSTRACT: This paper gives insight into detail design of the MPU Heavy Lifter. The innovative structure is developed for offshore heavy lifting operations like removing or installing platforms. Its hull is composed of highly reinforced and prestressed lightweight concrete forming slabs and walls with thicknesses between 0.3 m and 0.9 m. Information is provided on the design procedure, on the basis of design and on the design of B- and D-regions. Structural analysis is based on finite element analysis using shell elements. For design and verification of concrete sections the results are post-processed with specially developed software. For D-regions, truss and stress-field modelling supported by specialised finite element analysis is applied. The applicability of these design methods is verified with a series of small and large scale tests.

Tailor Made Concrete Structures – Walraven & Stoelhorst (eds)
© 2008 Taylor & Francis Group, London, ISBN 978-0-415-47535-8

Computer modeling and effective stiffness of concrete wall buildings

M.IJ. Schotanus & J.R. Maffei

Rutherford & Chekene Consulting Engineers, San Francisco, California, USA

ABSTRACT: Shake-table tests of a full-scale seven-story wall structure at the University of California at San Diego (UCSD) provide a crucial benchmark in evaluating methods that are currently being used to design mid-rise and high-rise concrete buildings in seismically active areas. The authors compare properties and characteristics of the UCSD test structure with twelve tall concrete core-wall buildings that have recently been designed for the western United States, and find that the test results are applicable to this type of structure. Using assumptions, methods, and software that are typical in design practice, the authors constructed linear and non-linear computer analysis models of the UCSD test structure. Iterations of assumptions for the linear models lead to recommended concrete stiffness properties, which are then compared to published recommendations that are often used in design. Recommended stiffness properties are lower than those commonly used in practice. Comparison of the non-linear models to test results shows a difficulty in matching building deformations while also matching overturning moments and shear forces. Both types of models show a significant influence of slabs engaging columns, and acting as outriggers, increasing overturning resistance and shear demand on the wall.

Tailor Made Concrete Structures – Walraven & Stoelhorst (eds)
© 2008 Taylor & Francis Group, London, ISBN 978-0-415-47535-8

Inelastic seismic response and damage analysis of a tall bridge pier

X. Zhu, J.-W. Huang & L.-Y. Song
School of Civil Engineering and Architecture, Beijing Jiaotong University, Beijing, China

ABSTRACT: The impulsive near-fault earthquake ground motions generally impose high demands on structures compared to ordinary ground motions. The seismic behavior of a reinforced concrete tall bridge pier is studied in this paper, using a realistic example of Huatupo bridge Pier #8. According to Park-Ang dual-parameter damage model, the maximum local damage index was obtained at the about height of 60 m, it is located at the medium to upper portion of this pier. The inelastic seismic response demands and seismic damage performance evaluation of reinforced concrete tall bridge pier subjected to pulse-type near-fault earthquake ground motions and corresponding equivalent pulses would be investigated. Except seismic behavior of tall bridge piers, this study was also intended to identify salient inelastic response and damage characteristics of tall bridge pier excited by near-fault ground motions, and to check whether pulse-type near-fault ground motions could be reasonably represented by simple equivalent pulses.

Tailor Made Concrete Structures – Walraven & Stoelhorst (eds)
© 2008 Taylor & Francis Group, London, ISBN 978-0-415-47535-8

Experimental investigation on the seismic behaviour of connections in precast structures

R. Felicetti, G. Toniolo & C.L. Zenti
Department of Structural Engineering, Politecnico di Milano, Milan, Italy

ABSTRACT: The paper is addressed to the identification of the behaviour parameters of a type of connection between floor or roof elements and the supporting beams. This identification will allow to give a correct representation of this type of joints within the overall numerical model for structural analysis and to give practical application to the criteria of capacity design. The first results of the experimental campaign are reported with reference to monotonic (push-over) and cyclic tests. Following the results of some preliminary trial tests, a complete framing of the experimental programme has been defined, covering the general testing protocol for all types of connections. Moreover, the guidelines have been drafted for a standardized quantification of the principal parameters of the connections seismic behaviour.

Tailor Made Concrete Structures – Walraven & Stoelhorst (eds)
© 2008 Taylor & Francis Group, London, ISBN 978-0-415-47535-8

Design approach for diaphragm action of roof decks in precast concrete building under earthquake

L. Ferrara & G. Toniolo

Department of Structural Engineering, Politecnico di Milano, Milan, Italy

ABSTRACT: The response of a structure to earthquake relies on the diaphragmatic behaviour of slabs at both storey and roof level. Cast-in-situ r/c slabs can easily behave as rigid diaphragm. The same holds for slabs built with precast elements and completed through in situ casting, provided peripheral ties and suitable reinforcement at connections with vertical elements is provided. Long span roofing of precast r/c industrial buildings generally consist of "panel elements" connected through mechanical devices, without any cast-in-situ completion, or even interposed with skylights. Diaphragm action can be still relied upon if the in plane rotational equilibrium of elements can be guaranteed by forces at peripheral connections: for this case a simplified design approach is proposed and validated in this paper.

Increasing the speed of construction

Tailor Made Concrete Structures – Walraven & Stoelhorst (eds)
© 2008 Taylor & Francis Group, London, ISBN 978-0-415-47535-8

Just in time mixture proportioning

Ken W. Day
Consultant

ABSTRACT: The paper describes a system for automatically designing the next truck of concrete in a few seconds, taking into account all relevant data available up to the time of batching. Such data should include current test data on concrete and constituent materials, anticipated concrete temperature, required slump and transport time. *This paper presents the view that, in order to achieve minimum variability, the contents of each truck of concrete need to be finally determined only a few minutes before that truck is batched, taking into account every piece of relevant information that can be made available at that point in time.*

Tailor Made Concrete Structures – Walraven & Stoelhorst (eds)
© 2008 Taylor & Francis Group, London, ISBN 978-0-415-47535-8

Dubai metro challenge for a fast track construction

Y. Gauthier, S. Montens, P. Arnaud & T. Paineau
Systra

ABSTRACT: Building a transportation infrastructure is a critical challenge for the fast growing city of Dubai. Both elevated and underground structures of the two first Metro lines of Dubai have been designed to be cost and time efficient. On the elevated section, the Railway viaduct is designed to be built extensively by precast techniques. The time efficient proven technique of precast segmental construction is implemented. The elevated station concept is in line with the standardization of the viaduct design and construction. At the end, the overall success of a speedy and timely construction depends on the quality of the planning: interface with Third Parties Projects, Utilities, resources, design freeze, quick decision process, are as usual, critical parameters.

Tailor Made Concrete Structures – Walraven & Stoelhorst (eds)
© 2008 Taylor & Francis Group, London, ISBN 978-0-415-47535-8

Balanced lift method – A new bridge construction method

S. Blail & J. Kollegger
Institute for Structural Engineering, Vienna University of Technology, Vienna, Austria

ABSTRACT: The so-called arch lowering construction method offers a possibility to build large arch bridges. As a counterpart to the lowering construction method the "balanced lift method" is presented in this paper. The main advantages of this new bridge construction method are savings in construction time and construction materials. It is suggested to build the bridge girders in vertical position and rotate them into the final horizontal position afterwards. The bridge girders can be built in combination with the pier using climbforming techniques which will allow a considerable cost reduction in manufacturing and speed up construction time. The range of the span length for the application of the balanced lift method lies between 50 m and 250 m.

Tailor Made Concrete Structures – Walraven & Stoelhorst (eds)
© 2008 Taylor & Francis Group, London, ISBN 978-0-415-47535-8

Fast track construction of 9.5 km long elevated expressway by largescale, prefabrication of superstructure

S. Sengupta
Span Consultants Pvt. Ltd., Bangalore, India

ABSTRACT: The traffic volume on existing national highway no. 7 from Bangalore city to Hosur city in southern India has substantially increased in recent years to an extent much beyond the capacity of the 4-lane highway. Apart from being a common stretch to the Golden Quadrilateral and North-South corridor of national highway network in India, this stretch provides daily access to large number of Information Technology and Bio-Technology professionals in Bangalore who commutes daily to the Electronic city appx. 9.5 km away from the downtown area. Due to the severe traffic congestion and consequent loss of precious man-hours on the road, the authorities had decided to construct an elevated expressway all along the road with full access control and tolling for providing a dedicated fast access on this corridor.

The four-lane superstructure takes off just after an existing flyover recently built and run straight into the Electronic City, Phase I & II approx. 9.5 km away. The elevated viaduct alongwith a multilevel interchange at the terminal point is being constructed on Build Operate & Transfer (BOT) basis using large – scale precast prestressed concrete (PSC) segmental, glued, matchcast technology and erected by Overhead Launching Girders without using any ground scaffold to avoid any traffic disturbance at GL. The four-lane superstructure is supported on single row of piers to be constructed along the central median of the highway. This is presently the longest flyover under construction in India. Fig. 1 shows a general view of the elevated expressway.

Tailor Made Concrete Structures – Walraven & Stoelhorst (eds)
© 2008 Taylor & Francis Group, London, ISBN 978-0-415-47535-8

An alternative tunnelling approach to accelerate urban underground excavation under water

F. Cavuoto
Structural & Geotechnical Designer, Naples, Italy

G. Colombo
Metropolitana Milanese S.p.A., Italy

F. Giannelli
Astaldi S.p.A., Italy

ABSTRACT: The paper presents a very safe, cost- and time-effective tunnelling approach contrived and applied to the staged excavation of a tall and wide underground opening (*cavern*) within the *Toledo Station*, as part of the new Subway *Line 1* in Naples. In this project the temporary support, usually provided by steel arches and shotcrete in aid to the frozen soil arch, was provided by the frozen soil arch alone by means of the Artificial Ground Freezing (AGF) technique until the completion of a first reinforced concrete liner following the excavation front (4 m behind), relying upon the high resistance of soil under freezing. In other words, the frozen soil arch acted both as structural and waterproofing element. With respect to traditional excavation technologies, the proposed solution proved to be very safe without penalizing time and economic aspects, plus allowing the thickness of the permanent support to be increased.

Poster Presentations

Life cycle design

Tailor Made Concrete Structures – Walraven & Stoelhorst (eds)
© 2008 Taylor & Francis Group, London, ISBN 978-0-415-47535-8

The influence of high slag dosages in cement on frost and de-icing salt resistance of concrete

S. Uzelac, A. Hranilović Trubić, I. Šustić & D. Würth
Civil Engineering Institute of Croatia, Zagreb, Croatia

ABSTRACT: New types of concrete („green" concrete, high and ultra high strength concrete) has developed recently. In meantime awareness of the need for reduction of greenhouse gas emissions in the construction industry has resulted in increasingly environmentally friendly cement production. It is known if the clinker part in cement is substituted with mineral additives (slag from a blast furnace, fly ash), the CO_2 emission is significantly reduced. The serviceability of a such cement type should be tested in local conditions with the aim of preserving effects of the environmentally friendly production.

Tailor Made Concrete Structures – Walraven & Stoelhorst (eds)
© 2008 Taylor & Francis Group, London, ISBN 978-0-415-47535-8

Case study: LCC analysis for Krk Bridge

I. Stipanović Oslaković
Civil Engineering Institute of Croatia, Zagreb, Croatia

D. Bjegović & J. Radić
Civil Engineering Institute of Croatia, Zagreb, Croatia
Faculty of Civil Engineering, University of Zagreb, Zagreb, Croatia

ABSTRACT: The Krk Bridge (constructed between 1976–1980) connects the mainland and the island of Krk, and consists of two arches including one of the largest conventionally reinforced concrete arch span in the world. During more than 25 years of service the bridge was exposed to very aggressive environment which caused corrosion problems. Therefore, structures of both arches are being repaired intensively during last 10 years. In this paper approximate costs of the maintenance and repair works performed during last 3 decades on the small and big arch of the Krk bridge are analysed. The actual costs are compared to the estimated costs of the life cycle management, if outer layer of carbon steel reinforcement had been replaced with stainless steel.

Tailor Made Concrete Structures – Walraven & Stoelhorst (eds)
© 2008 Taylor & Francis Group, London, ISBN 978-0-415-47535-8

Effect of GGBS additive on chloride ion binding of cements

K. Kopecskó & Gy. Balázs
Budapest University of Technology and Economics, Hungary

ABSTRACT: This paper deals with our test results on hardened and afterwards salt-treated cements modelling the influence of de-icing salt. Hydration both of naturally hardened and steam-cured cements were investigated as well as their chloride ion binding mechanisms by thermal tests and X-ray diffraction. The series of samples were cements with different amount of ground granulated slag (CEM I 42.5 N, CEM II/A 32.5 R, CEM III/A 32.5 N and CEM III/B 32.5 N-S). The chloride ion binding mechanism of ground granulated blast-furnace slag without cement clinker content was also studied. The formation of Friedel's salt and delayed ettringite increased between the ages 90 to 180 days. Test results indicated that steam cured cements can bind higher amount of chloride ions than naturally hardened ones. Chloride ion binding ability of tested cements in decreasing sequence was the following: (1) CEM III/B 32.5 N-S; (2) CEM III/A 32.5 N, (3) CEM II/B-S 32.5 R; (4) CEM I 42.5 N. It was also experimentally shown that ground granulated blast furnace slag itself is able to bind chloride ions.

Design strategies for the future

Tailor Made Concrete Structures – Walraven & Stoelhorst (eds)
© 2008 Taylor & Francis Group, London, ISBN 978-0-415-47535-8

Optimised Concrete Quality Control

K.W. Day
Consultant

ABSTRACT: The paper proposes that a more advanced form of Multigrade, Multivariable, Cusum QC be adopted in place of the limited "Families" approach in EN 206. The proposed system is easier to set up and operate and results in much earlier detection of problems and their causes and therefore in reduced variability, saving cement by enabling a smaller margin between mean and specified strengths.

Tailor Made Concrete Structures – Walraven & Stoelhorst (eds)
© 2008 Taylor & Francis Group, London, ISBN 978-0-415-47535-8

Full-scale test on a pile supported floor slab – steel fibre concrete only or in a combination with steel

J. Hedebratt
Tyréns AB (Building Consultants) and Department of Structural Design and Bridges,
Royal University of Technology, Stockholm, Sweden

J. Silfwerbrand
Swedish Cement and Concrete Institute, CBI and Department of Structural Design and Bridges,
Royal University of Technology, Stockholm, Sweden

ABSTRACT: The Ph. D project "Integrated Design and Construction of Industrial Floors" proceeds after the presentation of a Licentiate thesis covering methods to increase the quality of concrete floors, Hedebratt (2004). The aim for further studies is to develop directions for design and construction of pile supported steel fibre concrete, SFC floors. SFC is common in industrial floor slabs. In pile supported floor slabs also a combination of non-tensioned reinforcing bars and steel fibres have been used. Furthermore, neither Swedish nor European or any other known design guidelines cover steel fibres as the only reinforcement in pile supported floors or structural members. A common engineering advice is to disregard the ground support. The scope is to investigate the possibility to consider steel fibre only design solutions in a safe way and to compare it with a combined solution and as a reference performing full-scale tests and to develop design guidelines. The now ongoing test on a column supported deck emulates a half scale of an industrial supported floor slab but may also be considered to be full scale of a structural mushroom floor for small housing.

Monitoring and inspection

Tailor Made Concrete Structures – Walraven & Stoelhorst (eds)
© *2008 Taylor & Francis Group, London, ISBN 978-0-415-47535-8*

Damage assessment of fiber reinforced cement matrix under cyclic compressive load using acoustic emission technique

Yun Su Kim, SunWoo Kim & Hyun Do Yun
Department of Architectural Engineering, Chungnam National University, South Korea

ABSTRACT: In this study is devoted to the investigation of the Acoustic Emission (AE) signals in HPFRCC under monotonic and cyclic uniaxial compressive loading, and total four series were tested. The major experimental parameters include the type of fiber (PE and PVA), the hybrid type with steel cord and loading pattern. The test results showed that the damage progress by compressive behavior of the HPFRCC is characteristic for the hybrid fiber type and volume fraction. And from AE parameter value, it is found that the second and third compressive load cycle resulted in successive decrease of the amplitude as compared with the first compressive load cycle. Also, the Kaiser effect existed in HPFRCC specimens up to 80% of its ultimate strength. These observations suggested that the AE Kaiser effect has good potential to be used as a new tool to monitor the loading history of HPFRCC.

Tailor Made Concrete Structures – Walraven & Stoelhorst (eds)
© 2008 Taylor & Francis Group, London, ISBN 978-0-415-47535-8

Online-monitoring of concrete structures: Cost-effectiveness and application

Y. Schiegg
Swiss Society for Corrosion Protection (SGK), Zurich, Switzerland

L. Steiner & S. Rihs
Institute for Information Systems, Berne, Switzerland

ABSTRACT: The functionality and safety of reinforced concrete structures needs to be ensured over their entire service life. Since the reinforcement exhibits a relatively high risk for corrosion, these structures must be inspected regularly and observed damages have to be repaired. The control of reinforced concrete structures is typically performed by visual inspection. As soon as structural problems are observed, a more thorough and expensive analysis is performed. Online-monitoring could represent an alternative with higher accuracy of detection of damages and, thus, lower overall costs.

Diagnosis

Tailor Made Concrete Structures – Walraven & Stoelhorst (eds)
© *2008 Taylor & Francis Group, London, ISBN 978-0-415-47535-8*

Systematic condition investigation of concrete structures

J. Lahdensivu, S. Varjonen & M. Pentti
Department of Civil Engineering, Tampere University of Technology, Tampere, Finland

ABSTRACT: The aim of condition investigation of concrete structures is to provide the property owner and designers information to help them decide on the repair needs and possibilities of concrete structures and to select the repair methods best suited for each facility. The content of the condition investigation must be such that set goals are met. Generally, a condition investigation is made to determine the remaining service life of examined structures, their need of repair and safety. To achieve that, the investigation must reveal any damage to structures and defects in their performance. This requires determining the existence of damage, its extent, location, degree, impacts and future propagation by damage types.

Tailor Made Concrete Structures – Walraven & Stoelhorst (eds)
© 2008 Taylor & Francis Group, London, ISBN 978-0-415-47535-8

Seismic response of corroded r.c. structures

Anna Saetta & Paola Simioni
Department of Architectural Construction, University IUAV, Venezia, Italy

Luisa Berto & Renato Vitaliani
Department of Structural and Transportation Engineering, University of Padova, Padua, Italy

ABSTRACT: An accurate diagnosis of r.c. structures requires the investigation of their progressive degradation over time. As a matter of fact, the increasing damage resulting from the environmental attacks that the structure may suffer during its service life, affects not only the load bearing capacity, but also the failure mechanism, leading to a more brittle behavior. The loss of ductility strongly influences the structural response to external loads especially in seismic conditions and an effective non linear model able to account for these aspects is strongly required. In this paper, the preliminary results of an investigation concerning the effects of steel corrosion on the seismic response of r.c. structures are presented. Some case studies are analyzed under a moderate corrosive attack and the outcomes are discussed in terms of capacity curves and compared with the provisions of the European Code.

Tailor Made Concrete Structures – Walraven & Stoelhorst (eds)
© 2008 Taylor & Francis Group, London, ISBN 978-0-415-47535-8

Non destructive tests for existing R.C. structures assessment

S. Biondi & E. Candigliota
*Department of Design, Rehabilitation and Control of Architectural Structures, Pricos, University
"G. D'Annunzio", Chieti-Pescara, Italy*

ABSTRACT: In this paper in-situ non destructive tests for r.c. structure diagnosis are discussed. Three existing sport domes were analyzed in order to evaluate their global behavior and to design their seismic assessment. A wide program of in situ (non destructive) and laboratory (destructive) tests was carried out. Some literature proposals are used for test data interpretation and to validate non linear F.E.M. structural analyses.

Innovative materials

Tailor Made Concrete Structures – Walraven & Stoelhorst (eds)
© 2008 Taylor & Francis Group, London, ISBN 978-0-415-47535-8

The research on early age thermal cracking control of C50 HPC in main tower of Sutong Bridge

G.Z. Zhang, L.Q. Tu & S.K. Li
CCCG WuHan Harbour Engineering Design & research Co., Ltd, Hubei, China

ABSTRACT: In recent years, thermal crack at early age of concrete has been one of the restricted factors affecting the durability of concrete structure and becomes a hidden peril of the structures. Considering the characteristics of Sutong Bridge, such as high main tower, large mass of concrete, high strength grade of concrete, thermal crack at early stage of concrete was studied applying temperature-stress testing machine. When designing mix proportion of concrete, some methods were used, such as mixing a certain amount of fly ash, reducing dosage of cementitious materials, decreasing sand-aggregation ratio and mixing polypropylene fiber in the root of main tower. Early-age curing of concrete was also strengthened. The results indicated that good performance of C50 high performance concrete was got, including the workability, mechanical properties, the anti-cracking and volume stability. And there was no early age thermal crack in the project.

Tailor Made Concrete Structures – Walraven & Stoelhorst (eds)
© 2008 Taylor & Francis Group, London, ISBN 978-0-415-47535-8

Rock-fill concrete, a new type of concrete

M. Huang, X. An, H. Zhou & F. Jin
State Key Laboratory Hydroscience and Engineering, Tsinghua University, Beijing, China

ABSTRACT: Rock-fill concrete was first developed in China as an application of SCC in massive concrete construction, with lower cost and heat of hydration. Since then, various investigations have been carried out, showing that this type of concrete performs satisfactorily with regard to strength, permeability and other properties. It has been used in practical structures mostly as the mass concrete in China since Aug 2005 with the benefits of its low heat of hydration, short construction period as well as the low cost and environmental friendly.

Tailor Made Concrete Structures – Walraven & Stoelhorst (eds)
© 2008 Taylor & Francis Group, London, ISBN 978-0-415-47535-8

Structural concept, static and dynamic properties of RPC-BSS with high durability

Lanchao Jiang & Ri Gao
School of Civil Engineering & Architecture, Beijing Jiaotong University, Beijing, P.R. China

Liang Li
Central Research Institute of Building and Construction, NCC Group, Beijing, P.R. China

ABSTRACT: A new kind of beam string structure, named RPC-BSS, which consists of reactive powder concrete upper chord and cables, is presented in this paper. The material of RPC provides superior mechanical properties and high durability, and the cable provides huge tensile force. The structural concept is based on the combination of the advantages of RPC and beam string structural technology. Furthermore, static and dynamic properties of the structure are discussed, and the influence of different factors are taken into consideration in the analysis. The analytical results show that the new structure is of longer span, less weight, better aseismic characteristics and higher durability than conventional beam string structure.

Tailor Made Concrete Structures – Walraven & Stoelhorst (eds)
© 2008 Taylor & Francis Group, London, ISBN 978-0-415-47535-8

By-product materials in the production of next generation SCC

A.D. Kanellopoulos, M. Mathaiou, M.F. Petrou, I. Ioannou & M. Neophytou
University of Cyprus, Nicosia, Cyprus

ABSTRACT: Self-Compacting Concrete (SCC) is an advanced type of concrete that can flow under its own weight, completely filling formwork and achieving full compaction, even in the presence of congested reinforcement. In hardened state, SCC is dense, homogeneous and has similar engineering properties as the traditionally vibrated concrete. The current study examines the effect of various by-product materials on the development of high quality SCC mixtures.

Tailor Made Concrete Structures – Walraven & Stoelhorst (eds)
© 2008 Taylor & Francis Group, London, ISBN 978-0-415-47535-8

Low strength self compacting concrete for building application

Kosmas K. Sideris, Aggelos S. Georgiadis, Nikolaos S. Anagnostopoulos & Panagiota Manita
Laboratory of Building Materials, Democritus University of Thrace

ABSTRACT: Application of self compacting concrete for the construction of a new office building in the city of Xanthi, Greece is described in this paper. The building was especially designed to have a bioclimatic behaviour. Low strength (C25/30) and low fines self compacting concrete was designed for this application using a new mix design and quality control method developed at the Laboratory of Building Materials of Democritus University of Thrace (DUTH).

Tailor Made Concrete Structures – Walraven & Stoelhorst (eds)
© 2008 Taylor & Francis Group, London, ISBN 978-0-415-47535-8

Confirmation tests of integration of UFC forms left in place

M. Katagiri
Taiheiyo Cement Corporation Research & Development Center, Chiba, Japan

K. Kakida
Japan Railway Construction, Transport and Technology Agency, Hokuriku Shinkansen 2nd Construction Bureau, Toyama, Japan

T. Nihei
Railway Technical Research Institute, Tokyo, Japan

ABSTRACT: Certain parts of the construction site of Hokuriku Shinkansen in Niigata Prefecture are within 700 m of the coastline. Salt damage countermeasures are required for reinforced concrete structures. Accordingly, studies to reduce the covering of the concrete have been made by application of highly durable forms left in place made of UFC (ultra high strength fiber reinforced concrete). In this study, load tests were conducted to confirm that highly durable, thin-walled formwork left in place will not detach from the concrete. The tests used a 1/3 scale test piece. Test results showed no detachment, cracking, or decrease in rigidity under cyclic loading up to 2 million cycles. The highly durable forms left in place were integrated with the concrete, and the strain distribution along the vertical cross section at the center of test piece was approximately linear.

Tailor Made Concrete Structures – Walraven & Stoelhorst (eds)
© 2008 Taylor & Francis Group, London, ISBN 978-0-415-47535-8

Optimizing ready mix concrete for specific environmental conditions

J.A. Ortiz & M.E. Zermeño
Autonomous University of Aguascalientes, Aguascalientes, Mexico

A.C. Parapinski, A. Aguado & L. Agulló
Technical University of Catalonia, Barcelona, Spain

F.A. Alonso
Autonomous University of Chiapas, Chiapas, Mexico

ABSTRACT: Herein is presented a study of the effects of environmental conditions on the performance of concrete in both fresh and hardened states, and on the properties of the constituent materials of concrete, from an industrial perspective. The main objective was the optimization of concrete according to specific environmental conditions. The application of certain industrial procedures to ready mix concrete plants is proposed as a method to minimize the adverse effects of climate on concrete performance. Several experimental stages were first carried out to detect the influence of extreme environmental conditions on the thermal, workability and mechanical properties of concrete. The results indicate that aggregates, which are the predominant component of concrete, are especially labile to temperature. A basic methodology for optimizing concrete performance in specific environments was developed. This procedure has been successfully applied on an industrial scale in ready mix concrete plants, where it yielded significant savings in cement.

Tailor Made Concrete Structures – Walraven & Stoelhorst (eds)
© 2008 Taylor & Francis Group, London, ISBN 978-0-415-47535-8

Tailor made bridge design with Ultra-High-Performance Concretes

R.P.H. Vergoossen
ARCADIS, Rotterdam, The Netherlands

ABSTRACT: Since the early nineties different manufacturers can produce different concrete mixes with compressive strengths over 150 MPa. These mixes have not only very high strengths but there overall performance, such as resistance against aggressive agents, is much higher then for normal concrete. (i.e. strength classes C30/37 to C50/63) In the last years many studies and designs in the Netherlands and in the world have shown the great possibilities for these Concretes for bridge structures. Many types of bridges for various types of traffic and with various spans have been investigated or realised. Even existing bridges are strengthened or adapted to resist heavier loads or to expand there life cycle. In the paper different studies and designs are described. From these studies and designs some general conclusions can be drawn in relation to the possibilities for these new Ultra-High-Performance Concretes (UHPC) for there practical and cost effective use in structures.

Tailor Made Concrete Structures – Walraven & Stoelhorst (eds)
© 2008 Taylor & Francis Group, London, ISBN 978-0-415-47535-8

Development of Self-Consolidating Concrete in Hawaii using basalt aggregates

I.N. Robertson, G.P. Johnson & R. Ishisaka
University of Hawaii at Manoa, Honolulu, Hawaii, USA

ABSTRACT: The use of Self-Consolidating Concrete (SCC) has expanded rapidly over the past decade as a result of its many advantages over traditional concrete. The improved filling and passing ability, high quality finish and ease of placement often offset the increased initial cost of SCC. Past research on SCC has often recommended the use of rounded aggregate, such as river gravel, to improve flow characteristics. In many communities such aggregates are not available. In Hawaii, all concrete coarse aggregates, and much of the fine aggregates, are derived from crushing basalt rock. This can result in angular particles with poor aspect ratios. In addition, the high water absorption of the porous basalt aggregate can result in mix stability issues. In an effort to develop SCC mixtures for use in drilled shafts, a research study was initiated at the University of Hawaii. This paper provides an overview of this effort and highlights some of the difficulties encountered when using crushed basalt as the primary source of both coarse and fine aggregates.

Improved self-compacting concrete mixes for precast concrete industry

A. Ioani & O. Corbu
Technical University, Cluj-Napoca, Romania

H. Szilagyi
National Institute for Research in Construction (INCERC), Cluj-Napoca Branch, Romania

ABSTRACT: The paper presents a research program developed at The Technical University of Cluj-Napoca for self-compacting concrete (SCC) implementation into the Romanian precast concrete industry. The mixes were designed to fulfil the specific requirements of precast concrete: high early strength and high final strength for precast/prestressed applications, very good filling and passing ability, stability and segregation resistance, high-quality surface finishes. The experimental program parameters were: C50/60, C40/50, C30/37 strength classes, cement content (from 510 to 330 kg of cement/m^3), types and quantity of additions, types and quantity of admixtures (HRWRA, VMA), water amount, aggregate grading and the mix robustness. The concrete mixes properties in fresh and hardened state were presented and discussed.

Tailor Made Concrete Structures – Walraven & Stoelhorst (eds)
© 2008 Taylor & Francis Group, London, ISBN 978-0-415-47535-8

UHPFRC prestressed beams as an alternative to composite steel-concrete decks: The example of Pinel Bridge (France)

Thierry Thibaux

EIFFAGE TP – Neuilly sur Marne – France

ABSTRACT: This paper describes the design and construction of the third Pinel Bridge in France, in 2007. Located just to the west of Rouen, this small bridge has a single 27-m-long span and is 14-m wide. It will increase the traffic capacity of two existing bridges over three railroad tracks from two lanes to five. The bridge deck combines a conventional slab of standard reinforced concrete with seventeen parallel contiguous beams made of self-consolidating Ultra-High-Performance Fiber-Reinforced-Concrete (UHPFRC). The beams – designed by the contractor, Eiffage TP, as an alternative to a filler-beam deck – are prestressed with 28 internal strands placed in a very wide bottom flange

Tailor Made Concrete Structures – Walraven & Stoelhorst (eds)
© *2008 Taylor & Francis Group, London, ISBN 978-0-415-47535-8*

Ultra High Performance Concrete: Mix design and practical applications

N. Cauberg & J. Piérard
Belgian Building Research Institute, Brussels, Belgium

O. Remy
University of Brussels, Brussels, Belgium

ABSTRACT: Evaluating two interesting applications for UHPC, cladding panels and overlays, this project focused on some relevant aspects such as the mix design of UHPC, the shrinkage at early age, the fiber reinforcement and the flexural behaviour. As far as mix design concerns, the research optimized the choice of admixtures, (micro)fillers and the aggregate grading, obtaining a compressive strength between 125 and 180 N/mm^2 and excellent flexural behavior with the cocktail of micro- and macrofibers. Both restrained and unrestrained shrinkage have been evaluated, and the results seem not to limit the applications. Two practical applications have been studied and show the potential of this material: thin and large cladding panels with different types of reinforcement, together with new anchorage systems. Secondly, UHPC-overlays for old and new concrete elements seem to be an innovative solution for concrete surfaces exposed to wear or aggressive substances. Modeling and real-scale experiments have been compared for this application.

Tailor Made Concrete Structures – Walraven & Stoelhorst (eds)
© 2008 Taylor & Francis Group, London, ISBN 978-0-415-47535-8

Shear strength of reinforced concrete beams with recycled aggregates

Sang Kyu Ji, Won Suk Lee & Hyun Do Yun
Department of Architectural Engineering, Chungnam National University, South Korea

ABSTRACT: In this paper, three reinforced concrete beams using recycled aggregate were tested to evaluate their failure modes, shear behavior and shear strength of reinforced concrete beams. All specimens have a cross section of 170×270 mm and a shear span-to-depth of 2. Each specimen was simply supported and subjected to four-point loading. The results showed that the beams with recycled aggregates present the similar shear failure mode as the beam with natural aggregate. The codes are conservative and subsequently can be used for the shear design of reinforced concrete beams with recycled aggregates.

Codes for the future

Comparing EC2, ACI, and CSA shear provisions to test results

E.C. Bentz & M.P. Collins
University of Toronto, Canada

ABSTRACT: The shear provisions of the Canadian CSA code, the US ACI code and EC2 for members without shear reinforcement are compared to a database of 1601 experimentally observed shear failures. The conclusion is that the CSA shear provisions provide the best estimate of the observed shear strength, though some improvement could be made for strut-and-tie with high strength concrete. The ACI code, which has remained essentially unchanged for 40 years provides unconservative predictions both for some members controlled by the strut-and-tie provisions and for large members controlled by sectional shear failures primarily in the latter case due to the size effect being ignored. EC2 also provide poor matches to the observed results with the size effect being underestimated, and the influence of high reinforcement percentages being overestimated.

Review of European standards and guidelines for grouts for pre-stressed structures

J. Tritthart
Graz University of Technology, Graz, Austria

I. Stipanović Oslaković
Civil Engineering Institute of Croatia, Zagreb, Croatia

P.F.G. Banfill
Heriot Watt University, Edinburgh, UK

M. Serdar
University of Zagreb, Zagreb, Croatia

ABSTRACT: Within the research project "Improvement of properties of grouts for prestressing tendons and/or ground anchors" in the European COST Materials Action 534 "New Materials and Systems for Pre-stressed Concrete Structures" methods for testing rheology, stability of the grout suspension with respect to bleeding and settlement, setting time, expansion and mechanical strength of the grout were performed and evaluated. One of the aims of the project was to critically investigate which methods that are given in European standards and guidelines are indeed suited for testing grout in laboratory and on site. Some of the most interesting results and conclusions gathered during the project are presented in this paper.

Tailor Made Concrete Structures – Walraven & Stoelhorst (eds)
© 2008 Taylor & Francis Group, London, ISBN 978-0-415-47535-8

Biaxial tensile strength of concrete – Answers from statistics

L. Lemnitzer, L. Eckfeldt, A. Lindorf & M. Curbach
Institute of Concrete Structures, Technische Universität Dresden, Germany

ABSTRACT: Over the last century, numerous research projects investigating the multiaxial stress-strain-behaviour of plain concrete have been conducted. In most instances, the main focus of data evaluation has been on the compressive strength of concrete. This paper summarises and analyses the biaxial tensile strength data for plain concrete available from literature. The influence of transverse tensile stress on tensile strength of concrete is quantified.

Tailor Made Concrete Structures – Walraven & Stoelhorst (eds)
© 2008 Taylor & Francis Group, London, ISBN 978-0-415-47535-8

Measurements of the transmission length of pre-tensioned strands

C. Bosco & M. Taliano
Department of Structural and Geotechnical Engineering, Politecnico di Torino, Turin, Italy

ABSTRACT: This paper reports the main results of an experimental programme which was conducted to analyse a type of bond test for the evaluation of the bond characteristics of seven-wire low relaxation strands used in pre-tensioned structures. The experimental campaign highlights that the concrete strength marginally influences the transmission length.

Tailor Made Concrete Structures – Walraven & Stoelhorst (eds)
© 2008 Taylor & Francis Group, London, ISBN 978-0-415-47535-8

A hybrid RC-encased steel joist system

C. Amadio, L. Macorini & F. Patrono
Department of Civil and Environmental Engineering, University of Trieste, Trieste, Italy

G. Suraci
Structural Engineer in Udine, Italy

ABSTRACT: In the paper a hybrid reinforced concrete system, made of steel trusses encased in reinforced concrete members, is presented. Such a system can be used for both columns and beams either in cast-in-place or in precast constructions. The monotonic behavior up to collapse of hybrid beams and the cyclic behavior of beam-to-column connections have been investigated experimentally. The results achieved have showed as these composite members can guarantee advanced structural performance even when employed in dissipative frames in seismic regions.

Tailor Made Concrete Structures – Walraven & Stoelhorst (eds)
© 2008 Taylor & Francis Group, London, ISBN 978-0-415-47535-8

Strengthening of reinforced concrete roof girder with unbonded tendons cracking due to the exploitation

A.S. Seruga & D.H. Faustmann
Kraków University of Technology, Kraków, Poland

ABSTRACT: In the paper the practical possibility of strengthening of reinforced concrete roof girder with unbonded tendons is presented. The mention above girder, double T in cross section, has been cracked due to the external dead loading. During control of existing technical state it was located about 140 normal cracks in the bottom flange coming to the web as well as the diagonal cracks. The maximal width of normal cracks was equal to the 0.4 mm, and the diagonal cracks was equal to 0.6 mm. The system of strengthening is discussed. The influence of presstresing to structure deformation has been investigated. The readings of concrete strains at upper, bottom flanges and web girder have been taken with mechanical gauges DEMEC type. The results obtained from investigations were compared to the ones from FEM numerical analysis. The final findings dealing with the applied technology of strengthening has been also discussed.

Tailor Made Concrete Structures – Walraven & Stoelhorst (eds)
© *2008 Taylor & Francis Group, London, ISBN 978-0-415-47535-8*

Development of field data for effective implementation of the mechanistic empirical pavement design procedure

N. Ala
Civil Engineering, University of Nebraska-Lincoln, Lincoln, NE, USA

M.A. Stanigzai
NaBRO-UNL, Lincoln, NE, USA

A. Azizinamini
Civil Engineering, University of Nebraska-Lincoln, Lincoln, NE, USA

M. Jamshidi
Nebraska Department of Roads, Lincoln, NE, USA

ABSTRACT: Implementation of the Mechanistic Empirical (M-E) Pavement Design procedure in Nebraska is being progressed. M-E design is a new way of designing pavements and predicting their long term performance. The approach is composed of two components. Engineers have dealt with the mechanistic component for many years. However, complexities arise from the fact that pavements are subject to accumulative damages due to traffic loads and environmental factors. The empirical component is predicting accumulative damage using damage models obtained through experimentation and monitoring. FHWA has developed the M-E Pavement Design Guide (MEPDG) software to aid in the implementation of the new M-E design procedures. Damage models used in the software have been calibrated based on Long Term Pavement Performance (LTPP) data obtained from all over the U.S. which could result in an inaccurate prediction in any given State. The objective is to develop data to calibrate the existing MEPDG models so that its damage predictions would be more accurate for the way we build and maintain pavements in the State of Nebraska. Sensitivity analysis helps to validate the reasonableness of the model predictions, to identify problems in the soft-ware. Results from all the simulations showed that almost all of the cases produce reasonable values for transverse cracking, faulting, and IRI.

Tailor Made Concrete Structures – Walraven & Stoelhorst (eds)
© 2008 Taylor & Francis Group, London, ISBN 978-0-415-47535-8

The effect of wire drawing lubricant residues on the bond characteristics of prestressing strand

A. Osborn
Wiss, Janney, Elstner Associates Inc., New York, NY, USA

J. Connolly & J. Lawler
Wiss, Janney, Elstner Associates Inc., Northbrook, IL, USA

ABSTRACT: WJE performed a four-year study of US produced prestressing strand aimed at determining why its bond properties have declined over the past 20 years. This study was funded by the National Cooperative Highway Research Program (NCHRP 10-62). WJE found that only certain manufacturers were making strand with deficient bond as a result of excess residual wire drawing lubricants and other factors. Surface tests were developed by the research that can be used to predict the bond properties of the strand. These tests were compared to transfer length tests and pullout tests of untensioned strand embedded in concrete and mortar. The tests developed by WJE are proposed for use in strand production facilities to qualify the bond properties of strand before it is shipped to customers.

Tailor Made Concrete Structures – Walraven & Stoelhorst (eds)
© 2008 Taylor & Francis Group, London, ISBN 978-0-415-47535-8

Wind loading parameters – Measurements vs. Croatian standard

Ana Krecak & Petar Sesar
Civil Engineering Institute of Croatia, Zagreb, Croatia

ABSTRACT: New Croatian Standard assesses wind loads to bc used in the structural analyses of buildings up to height of 200 m, bridges up to a span of 200 m and footbridges up to a span of 30 m. According to wind loading map teritory of Croatia is divided in five zones, and mean reference wind speed for the fifth zone equals 50 m/s. This is extreme value, especially when other parameters (i.e. exposure coefficient) are taken into account. Eventually this can lead to increase of wind loading three or four times compared to the standards recently used. Intention of this paper is to discuss some issues which arise during project of wind protection of one motorway section in Croatia which is exposed to extreme wind loading.

Dynamic measurements of wind speed along mentioned motorway section are under way within scientific research of wind protection structures which are already built.

Tailor Made Concrete Structures – Walraven & Stoelhorst (eds)
© 2008 Taylor & Francis Group, London, ISBN 978-0-415-47535-8

Post-tensioned slab on ground

C.K. Cheong
Special Projects Design Manager, VSL Australia Pty Ltd

ABSTRACT: Introduce design concept of post-tensioned slabs on ground and their use as pavements for Distribution Centers.

Architectural concrete

Tailor Made Concrete Structures – Walraven & Stoelhorst (eds)
© 2008 Taylor & Francis Group, London, ISBN 978-0-415-47535-8

NSP Arnhem Central Transfer hall

M. de Boer, J.L. Coenders, P. Moerland, S. Hofman & J.C. Paul
Arup, Amsterdam, The Netherlands

ABSTRACT: This paper discusses the current design status of the NSP Arnhem Central Transfer hall project in Arnhem, The Netherlands. The project involves a complex geometrical, double-curved concrete shell roof design by UN Studio architects for a project in which many functions are combined, the Transfer hall being a merging point of passenger, commercial and social interchanges, a multi-use development integrating programme and flows of people and vehicles. The paper will introduce the project and will discuss the design methods applied to convert a complex geometrical vision to a feasible structural design.

Developing a modern infrastructure

Tailor Made Concrete Structures – Walraven & Stoelhorst (eds)
© 2008 Taylor & Francis Group, London, ISBN 978-0-415-47535-8

Design of two curve cable stayed bridges with overlapping decks supported by a single "X" shape tower

Catão Francisco Ribeiro & Cláudio Watanabe
ENESCIL Engenharia de Projetos, São Paulo, Brazil

Heitor Afonso Nogueira
ANTW Engenharia de Projetos, São Paulo, Brazil

ABSTRACT: The objective of this paper is to describe the Real Park Complex bridge located in the city of São Paulo, Brazil. This structure, conceived by the São Paulo Municipality City Hall and implemented by EMURB (Empresa Municipal de Urbanização – Municipality Urbanization Company), is aimed at connecting the Jornalista Roberto Marinho Avenue with the Nações Unidas Avenue and the Marginal do rio Pinheiros Avenue, both of which are important city thoroughfares.

Tailor Made Concrete Structures – Walraven & Stoelhorst (eds)
© 2008 Taylor & Francis Group, London, ISBN 978-0-415-47535-8

Cable-stayed bridge over the Labe at Nymburk – Hybrid structure tailored for simple construction

M. Kalný, V. Kvasnička & P. Němec
Pontex Consulting Engineers Ltd, Prague, Czech Republic

A. Brnušák
SMP CZ, a.s., Prague, Czech Republic

ABSTRACT: The cable-stayed bridges are competitive for the spans over 100 m to other bridge technologies due to small depth of superstructure and diversity, which enable their individual forming and matching to the given natural or urban environment. The cable-stayed bridge over the Labe River at Nymburk is the first bridge in the Czech Republic of extradosed type with low slender pylons and two lateral suspension planes of stays. The main span of 132 m has 52 m long composite part in the middle, which was floated on barges and then lifted into final position. The technology of construction was very straightforward, however from the beginning a great attention was given to appearance and harmony between the bridge and its location.

Tailor Made Concrete Structures – Walraven & Stoelhorst (eds)
© 2008 Taylor & Francis Group, London, ISBN 978-0-415-47535-8

Ultimate limit state analysis of a segmented tunnel lining

A.J.T. Luttikholt
Witteveen+Bos, Amsterdam, The Netherlands

A.H.J.M. Vervuurt
TNO Built Environment and Geosciences, Delft, The Netherlands

J.A. den Uijl
Delft University of Technology, Delft, The Netherlands

ABSTRACT: In order to gain a better understanding of the complex structural lining behaviour of bored tunnels in the soft Dutch soil, a test set-up has been developed at Delft University of Technology. In this facility a full-scale model has been subjected to construction and serviceability loading conditions. Finally, two tests have been performed in which the specimen was loaded until failure. Simulations with the FE program DIANA were conducted for studying the lining behaviour in more detail and validating numerical models.

Tailor Made Concrete Structures – Walraven & Stoelhorst (eds)
© 2008 Taylor & Francis Group, London, ISBN 978-0-415-47535-8

Conceptual design of offshore concrete structures

T.O. Olsen

Dr.techn. Olav Olsen a.s, Norway

ABSTRACT: There are remarkable offshore concrete structures. How they are conceptually developed? What are the functional requirements that shape these structures, and make them competitive?

Tailor Made Concrete Structures – Walraven & Stoelhorst (eds)
© 2008 Taylor & Francis Group, London, ISBN 978-0-415-47535-8

Design and construction of DELTA CITY shopping mall concrete structure in Belgrade

S. Marinkovic, V. Kokovic & I. Ignjatovic
Faculty of Civil Engineering, University of Belgrade, Serbia

V. Alendar
DNec, Belgrade, Serbia

ABSTRACT: The structure of DELTA CITY shopping mall in Belgrade consists of two separated structures: the structure of the mall and the structure of the multi-story open garage. The overall dimensions of the irregular layout of the structure are 210 m × 110 m, with four main levels in the mall and five parking levels in the garage. Because of the different exposure conditions, the structure of the garage is separated from the mall's structure with an expansion joint at all the levels, except at the level of the foundation slab. The mall's structure is designed without any expansion joints, except for temporary joints during the construction stage. It consists mainly of reinforced concrete cantilevered columns cast in place, at typical spans of 8.2 m × 8.2 m, precast reinforced concrete simple beams and precast prestressed hollow core slabs. The reinforced concrete frames cast in place are designed at the facades and at few locations in the interior, to provide additional seismic resistance. The main structure of the open multi-story garage consists of reinforced concrete frames and precast hollow core slabs and it is designed for exposure class XD3, with special attention paying to durability conditions. Due to mainly precast structure, especially hollow core slabs as a floor solution, the complete concrete structure of approximately 80,000 m^2 is constructed for less then 12 months.

Tailor Made Concrete Structures – Walraven & Stoelhorst (eds)
© 2008 Taylor & Francis Group, London, ISBN 978-0-415-47535-8

Small concrete pedestrian bridge with integral abutments – An alternative solution for pedestrian bridges over highways

Andreas Keil & Sandra Hagenmeyer
Schlaich Bergermann und Partner, Stuttgart, Germany

Jens Schneider
Frankfurt University of Applied Sciences, Department of Architecture and Civil Engineering, Frankfurt, Germany

ABSTRACT: Although modern computer calculation methods allow checking upper and lower boundary conditions of a structure – such as influence of different states of earth pressure on abutment walls - quickly and easily, most bridges are still planned with joints at abutments to ensure an easy planning and to be conform with German road authority regulations. This is inefficient for smaller bridges as joints at abutments and their maintenance are of significant costs looked upon from a life-cycle point of view. This paper describes the elegant solution for a small, 2 field pedestrian bridge (length 37,10 m) with integral abutments over a 4-lane highway that could serve as a model for highway bridges and replace the regular solutions - often heavy and clumsy – for such structures usually used in Germany. A simple structural solution with a soft sandwich panel behind the end walls and the earth and a stabilized earth-fill dam allows the deformation of the bridge deck for temperature loads. Calculations with different earth pressure distributions behind the walls proved the serviceability without joints for winter and summer conditions. The Y-shaped middle column adds to the architectural concept of a minimalized and tailor made concrete structure to make concrete structures more positively accepted in public.

Tailor Made Concrete Structures – Walraven & Stoelhorst (eds)
© 2008 Taylor & Francis Group, London, ISBN 978-0-415-47535-8

"Ravine des trois bassins" Bridge in La Reunion Island. A successful application of extradosed prestressing

Patrick Charlon
EIFFAGE TP, Neuilly sur Marne, France

Jacques Frappart
ARCADIS

Designing structures against extreme loads

Tailor Made Concrete Structures – Walraven & Stoelhorst (eds)
© 2008 Taylor & Francis Group, London, ISBN 978-0-415-47535-8

Seismic behaviour of precast column to foundation joint

G. Metelli

University of Brescia, Brescia, Italy

P. Riva

University of Bergamo, Bergamo, Italy

ABSTRACT: This paper aims at presenting the results of two experimental tests concerning the cyclic behaviour of a dissipative column-to-foundation connection for precast concrete elements. The joint is characterized by the use of high strength threaded steel bars in grouted sleeves and by steel support elements which allow an easy assemblage and centering of the column, meanwhile ensuring its stability before the grout injection. The tests allow to compare the response of cast-in-place connections against pocket foundation and grouted sleeve solutions.

Tailor Made Concrete Structures – Walraven & Stoelhorst (eds)
© 2008 Taylor & Francis Group, London, ISBN 978-0-415-47535-8

Probabilistic correlation of damage and seismic demand in R/C structures

M. Botta
Granted researcher, Regione Umbria, Italy

M. Mezzi
Dept. of Civil and Environmental Engineering. University of Perugia, Perugia, Italy

ABSTRACT: Within a probabilistic performance-based evaluation procedure, the influence of the geometrical and mechanical uncertainties on the definition of the demand parameters correlated to the damage states, is assessed. Numerical simulations are carried out on a r/c sample column, leading to the definition of fragility curves of the story drift ratio for different basic damage conditions.

Tailor Made Concrete Structures – Walraven & Stoelhorst (eds)
© 2008 Taylor & Francis Group, London, ISBN 978-0-415-47535-8

Seismic response of coupled wall-frame structures on pile foundations

S. Carbonari

Department of Architecture, Constructions, Structures, Università Politecnica delle Marche, Ancona, Italy

F. Dezi

Department of Materials and Environment Engineering and Physics, Università Politecnica delle Marche, Ancona, Italy

G. Leoni

Department ProCAm, Università di Camerino, Ascoli Piceno, Italy

ABSTRACT: A study on the seismic behaviour of coupled wall-frame structures founded on piles is presented. Three different foundation soils are considered and the seismic excitation at the outcropping bedrock is represented using 7 real accelerograms. Local site response analyses are performed to evaluate the incoming free-field motions. A numerical model, accounting for pile-soil-pile interaction and for material and radiation damping, is used to evaluate the impedance matrix and the foundation input motion. Dynamic analyses are then performed in the time domain by considering fixed (FBMs) and compliance-base models (SSIMs). Applications show that the effects associated to the rocking behaviour of the wall sensibly affect the dynamic response of the system increasing the structural displacements and producing a change in the distribution of the seismic demand within the wall and the frame.

Tailor Made Concrete Structures – Walraven & Stoelhorst (eds)
© *2008 Taylor & Francis Group, London, ISBN 978-0-415-47535-8*

Displacement based design of BRB for the seismic protection of R.C. frames

T. Albanesi, A.V. Bergami & C. Nuti
Department of Structures, University of Roma Tre, Rome, Italy

ABSTRACT: Different rehabilitation systems have been developed to upgrade the seismic performance of existing reinforced concrete frame buildings with non-ductile detailing: in particular, buckling restrained dissipative steel braces (BRB) offer many advantages. In this paper a displacement-based procedure to design dissipative BRB for the seismic protection of masonry-infilled frames is proposed. A two-fold performance objective is considered to protect the structure against the collapse and the non-structural damage by limiting global displacements and interstorey drifts so that structural and infill integrity is granted under a given seismic event. Positioning these devices in a structure to maximize their effectiveness at minimum cost is a very important issue which is considered too. As an example, an infilled-frame building, designed according to the non-seismic Italian Code and thus only for vertical loads, is analyzed. Non linear static analyses to assess the effectiveness of the proposed rehabilitation design procedure are performed.

Tailor Made Concrete Structures – Walraven & Stoelhorst (eds)
© 2008 Taylor & Francis Group, London, ISBN 978-0-415-47535-8

Seismic behavior of residential concrete walls

Sergio M. Alcocer, Alfredo Sánchez-Alejandre & Julián Carrillo
Instituto de Ingeniería, UNAM, Mexico

Roberto Uribe & Ángel Ponce
Center of Technology, CEMEX, México

ABSTRACT: New economic housing in Mexico is mainly constructed using low-rise reinforced concrete wall structures. Walls are thin (75- to 100-mm wide) and lightly reinforced, made of either normalweight, cellular or self-compacting concrete. Most walls are cast-in-place. Due to these characteristics, current seismic resistant design provisions are not applicable. To develop analysis, design and construction guidelines a research program has been underway over the past 3 years. Thirty three full-scale reinforced concrete walls have been tested under lateral loads, while maintaining constant vertical stresses equivalent to 0.25 MPa. Experimental variables have included type of loading (monotonic vs. cyclic), type of concrete (normalweight, cellular and self-compacting), aspect ratio, opening size, type of reinforcement (steel bars with $f_y = 420$ MPa and welded-wire mesh with $f_y = 500$ MPa) and the amount of wall shear reinforcement (0, $\approx$50 and $\approx$100 percent of minimum wall reinforcement ratio as required in ACI 318-05, i.e. 0.0025). Welded-wire mesh is characterized by low ductility ($\varepsilon_{su} \approx 3\%$).

The aim of the paper is to discuss experimental results and to compare them with several analytical models available in the literature and with the model developed for this study.

Tailor Made Concrete Structures – Walraven & Stoelhorst (eds)
© 2008 Taylor & Francis Group, London, ISBN 978-0-415-47535-8

Modification of the concrete properties after fire

György L. Balázs
Department of Construction Materials and Engineering Geology Budapest
University of Technology and Economics

Éva Lublóy
Ph.D. Student

ABSTRACT: Fast development of construction technology requires new materials. Great number of experiments supported that in the case of concretes with quartz gravel aggregate content, the spalling of concrete cover can be decreased by applying polypropylene fibres (Janson, Boström, 2004). In the case of lightweight aggregate concrete the utilisation of synthetic fibres could also decrease the possibility of the spalling of the concrete cover. It has been observed that the compressive strength changed differently in the case of normal weight or lightweight concrete (Faust, 2003). The residual compressive strength of concrete exposed to high temperatures is influenced by the type of applied cement, type of applied aggregate and type of applied fibres. Test results will be presented indicating ranges of influences on reduction of strength.

Tailor Made Concrete Structures – Walraven & Stoelhorst (eds)
© 2008 Taylor & Francis Group, London, ISBN 978-0-415-47535-8

State-of-the-art structural designs and axial shortening studies of super high columns in a tall building

P. Boonlualoah & S. Boonlualoah
PBL Group Limited, Bangkok, Thailand

ABSTRACT: An under construction 28-storey reinforced concrete building was instrumented to observe its axial shortening behavior. A number of columns and core walls at 1*st* floor levels have been instrumented for axial shortening measurements. A brief outlines on the field experiment, the laboratory concrete testing program, and the method used to estimate long-term axial shortening is presented.

Tailor Made Concrete Structures – Walraven & Stoelhorst (eds)
© 2008 Taylor & Francis Group, London, ISBN 978-0-415-47535-8

Results of an experimental research on precast structures under seismic actions

F. Biondini & G. Toniolo

Department of Structural Engineering, Politecnico di Milano, Milan, Italy

ABSTRACT: This paper investigates the seismic performance of one-storey reinforced concrete structures for industrial buildings. To this aim, pseudodynamic tests on full-scale structures have been carried out. Two structural prototypes consisting of six-columns with two lines of beams and roof elements placed side by side have been designed. The prototypes differ each from the other only for the orientation of the roof elements with respect to the direction of the seismic action. The results of these tests show that an effective horizontal diaphragm action can be activated even if the roof elements are not connected among them, and confirm the expected good seismic performance of these precast systems.

Increasing the speed of construction

Tailor Made Concrete Structures – Walraven & Stoelhorst (eds)
© 2008 Taylor & Francis Group, London, ISBN 978-0-415-47535-8

Incremental launching of final composite bridge deck

L. Sasek
Mott MacDonald Prague, Czech Republic

ABSTRACT: Launching of final composite bridge deck has all advantages of launching method such as erection speed, lower costs and high quality of the bridge. Further innovation for composite bridges is the position of sliding bearings and the sliding trace structure.

Tailor Made Concrete Structures – Walraven & Stoelhorst (eds)
© 2008 Taylor & Francis Group, London, ISBN 978-0-415-47535-8

Shear load capacity of concrete slabs with embedded ducts

J. Schnell & C. Thiele
Technical University of Kaiserslautern, Germany

ABSTRACT: In the context of a research scheme supported by the German Concrete and Civil Engineering Organization and the German Institute for Civil Engineering investigations were carried out on reinforced concrete slabs provided with embedded ducts. Even related to small dimensions of block-outs associated with such embedded ducting considerable reductions of load-bearing capacity compared to the load-bearing capacity of a floor slab not comprising ducting [1] may occur.

Tailor Made Concrete Structures – Walraven & Stoelhorst (eds)
© 2008 Taylor & Francis Group, London, ISBN 978-0-415-47535-8

Construction of low Langer bridges

S.Yoda, M. Ooba, M. Koga & S. Suzuki
East Japan Railway company, Tokyo, Japan

ABSTRACT: We reconstructed three railway bridges for four rail routes passing over a road to be widened. We had only 10 hours for the work time because of the traffic conditions. Therefore we had to take the way that made it possible to construct the bridges in a short period. The way was that we constructed Langer bridges at the place nearby and slid along the abutments. We also considered aesthetic structural form and adopted the Langer bridge which has a low rise ratio of 1:9.

Tailor Made Concrete Structures – Walraven & Stoelhorst (eds)
© 2008 Taylor & Francis Group, London, ISBN 978-0-415-47535-8

Rapid construction of long span precast concrete box girders for Incheon Bridge viaducts constructed with FSLM

K.Y. Choi, D.O. Kang, K.L. Park, C.H. Lee, H.Y. Shin & M.G. Yoon
Samsung Corporation, Seoul, Korea

ABSTRACT: The FSLM (Full Span Launching Method) is being successfully applied to the viaducts of the Incheon Bridge. This method made it possible to construct the long marine viaducts over 8 kilometers within 3 years. The integrated monitoring of a girder with 50 m span was conducted through the whole construction sequences. It verified the applicability and efficiency to the rapid construction of marine bridges with FSLM.

Author Index